Dust Explosions

Dust Explosions

Jean Cross

Commonwealth Scientific and Industrial Research Organization
Division of Fossil Fuels
Sydney, Australia
Formerly with
Wolfson Electrostatics Advisory Unit
Southampton, University
Southampton, United Kingdom

and

Donald Farrer

Wolfson Electrostatics Advisory Unit
Southamptom University
Southamptom, United Kingdom

PLENUM PRESS • NEW YORK AND LONDON

Library of Congress Cataloging in Publication Data

Cross, Jean, 1947–
Dust explosions.

Bibliography: p.
Includes index.
1. Chemical plants—Safety measures. 2. Dust explosion. I. Farrer, Donald. II. Title.
TP155.5.C7 660.2'804 82-7499
ISBN 978-1-4615-6871-1 ISBN 978-1-4615-6869-8 (eBook) AACR2
DOI 10.1007/978-1-4615-6869-8

Softcover reprint of the hardcover 1st edition 1982
A Division of Plenum Publishing Corporation
233 Spring Street, New York, N.Y. 10013

Preface

The interest of the media in dust explosions increased considerably following two major grain-elevator disasters in the United States in 1979. However, these were not isolated incidents and were statistically unusual only in the high loss of life involved.

Any oxidizable material that is dispersed in fine powder form may be explosive, and ignition sources with sufficient energy to ignite a dust cloud are easily produced in normal industrial processing. Dust fires and minor incidents are not uncommon in many industries, but fortunately the combination of events and circumstances that must coincide for a large-scale explosion arise only rarely. Nevertheless, this is often more by luck than by good management and many potentially hazardous situations are common in industry.

An explosive dust cloud and the circumstances in which it can ignite are not as simple to define as the equivalent situation in gases or flammable vapors. A large number of definitions and experimental tests have been devised to characterize the explosibility of dusts and ignition sources. The aim of this book is to provide a guide describing conditions in industry that could lead to dust explosions and the means to avoid them. Ignition sources and the way in which they can arise in powder processing are discussed and illustrated by case histories of reported incidents. The methods by which the potential hazards of a process or product can be evaluated are described, with special attention paid to the interpretation of the results of the different experimental methods. Finally, the commonly quoted ignition characteristics of powders and their relevance to the industrial situation are evaluated.

Wolfson Electrostatics Unit
Southampton University

Jean Cross*
Donald Farrer

*Now CSIRO, Sydney, Australia.

Contents

Notation

a	radius of a particle
c	mass concentration of dust
d	distance or length
f	vent ratio = vent area/vessel volume
k	cell constant
l	length
m	mass
p	pressure
q	charge
r	distance (polar coordinates)
t	time
v	velocity
x	distance (cartesian coordinates)
A	area
C	capacitance
D	density of material
E	electric field
F	force
I	current
K	chemical reaction rate
K_{st}	dust explosion classification
L	inductance
M, N	constants
P	powder cloud density
Q	charge density
R	electrical resistance
S	surface area
T	temperature
V	voltage (or volume of vessel)
W	spark energy
R	chemical gas constant
ϵ_0	permittivity of free space = 8.8×10^{-12} f/m
ϵ	dielectric constant
ρ	resistivity
σ	conductivity
τ	time constant
γ_c	constant
η	viscosity
κ	reaction rate
ω	frequency

1

Introduction

1.1. HOW, WHERE, AND WHY DUST IGNITIONS OCCUR

Any material that will burn in air when it is in a solid form may explode when it is in the form of a finely divided powder. Even materials that oxidize more slowly than would normally be implied by the term "burning" can ignite catastrophically if the particle size is small. Explosions of foods pharmaceuticals, grain products, organic materials, polymers, and metals all occur. Oxidation is an exothermic reaction. Normally in a solid sample the heat formed is easily absorbed into the solid. In a powder, however, the surface area on which oxidation occurs is very large and the volume of the particle is very small so the temperature rises. This increases the rate of oxidation, creating yet more heat, so a runaway situation is rapidly reached.

There is some evidence that solids vaporize before exploding and that the production of sufficient heat to vaporize the surface is a necessary part of the explosion. It is not known whether this is true for all materials, although it certainly plays an important part in the ignition of coal dust, where methane gas is given off, and in the burning of many polymers that partially decompose as ignition begins.

When a powder is settled into a layer or heap, or when the dust cloud is very dense, there may be insufficient oxygen to allow the reaction to proceed rapidly enough for an explosion. For most dusts there is an optimum density of particles in the air for the propagation of a dust explosion: the particles should be close enough together for the heat of one particle to initiate reaction in the next, but far enough apart to allow free access of oxygen. This leads to the concept of a *minimum* and *maximum explosion concentration*.

The maximum explosion concentration is not a well-defined parameter and is seldom measured. From a safety viewpoint it is dangerous to assume that powder above a certain density is safe because dust layers can burn or smoulder slowly and can also self-heat to ignition. Dust layers are also very easily disturbed to create an explosive cloud. It is often difficult to detect the slow burning of settled dust, and this can provide a particularly dangerous ignition source when it is moved to another part of the system or disturbed into a cloud.

The minimum explosive concentration (which, following the terminology of gases and vapors, is sometimes known as the *lower explosive limit* or LEL) is better defined and is one of the parameters by which an explosive dust is characterized.

Dust explosions can occur whenever there is a combination of a dust cloud in the air within the explosive concentration range and an ignition source. Ignition sources are provided by electrical or electrostatic sparks, hot surfaces, overheated powder particles, or any other source of sufficient energy to initiate reaction in a few particles. There are so many ignition sources that can arise under fault conditions or from operator error that it is seldom realistic to base a safety policy entirely on the elimination of ignition sources. If an explosive cloud might occur in an area, it is necessary to ensure either that explosion is prevented by reducing the oxygen level in the area or that the results of an explosion are controlled by venting, suppressing, or containing the pressure rise.

It cannot be over emphasized that, in the majority of major dust explosions, the greatest damage occurs during secondary explosions. The primary explosion is often limited to a small-scale incident in a limited area but this then forms a high-energy ignition source for dust in other areas or for dust clouds raised by the initial explosion.

In designing equipment and plants to eliminate hazards and minimize the damage due to failure, it is necessary to consider not only each section individually but also the plant as a whole: There must be no dust layer to aid propagation of an explosion from one area to another, and a fault in one area must not produce a hazardous condition in the next operation. It is not uncommon for the consequences of a small ignition in one area to initiate an explosion in another area because of bad positioning of vents or lack of foresight concerning the path that would be taken by debris or burning powder.

1.2. IGNITION SOURCES—STATISTICAL DATA

Clear statistics concerning the causes of fires and explosions involving dusts are not readily available for most industries. Data published by organizations in different countries do not agree since the sources do not deal with data on the same basis. Different industries may well have different problems and this reflects on the most common ignition source attributed to each industry. However, it is also possible that once a source has been identified, accidents become wrongly attributed. Thus one particular source becomes "fashionable" in a particular industry with little justification in fact.

The majority of accidents widely reported, and for which a major study is carried out, are incidents where there is considerable damage. In these cases reliable identification of the source of ignition is difficult. Minor fires and explosions, some of which could easily have led to severe accidents in only slightly different circumstances, seldom reach the statistical reports. Often, in the interest of a rapid return to full production, a complete investigation of causes is not carried out.

Although it is possible to gain considerable knowledge of the ignition source from analysis of the results of an explosion—such as the fire damage and the range of debris—this exercise is seldom carried out except where damage is too severe to separate which faults caused the explosion and which were caused by it.

Often, the study reveals two or three possible ignition sources and the accident is reported as an unknown source, or a guess will be made as to the most probable source given the various possibilities.

The result is that it is only possible to use the published statistics and case histories as a general guide to possible ignition sources, plant failures, and situations that should be avoided.

Reports of fires and explosions are published by the *Fire Protection Journal* in the United Kingdom, The National Fire Protection Association in the United States, the Arbeitsschutz in Europe, and also by organizations representing insurance companies and major chemical companies.

Although the published case histories cannot be considered as a good record of the actual causes of fires and explosions in industry, they are an extremely useful stimulation to the imagination when considering all the possible problems that could occur in a particular installation. In particular, they often clearly show up the interactions between different parts of the plant and the reasons that caused a small incident to escalate into a major disaster. For this reason a number of case histories and examples are given throughout this book. They cannot be considered a representative sample of industrial accidents—simply examples of what may occur and illustrations of different types of incidents.

Two accidents occurred in the grain industry in the United States in December 1977. They resulted in a major reawakening of interest in the problems of dust explosions because they were very close together in time and resulted in a large loss of life. The following extracts of reports on the incidents are taken from the *Fire Journal* of September 1978 published by the National Fire Protection Association.

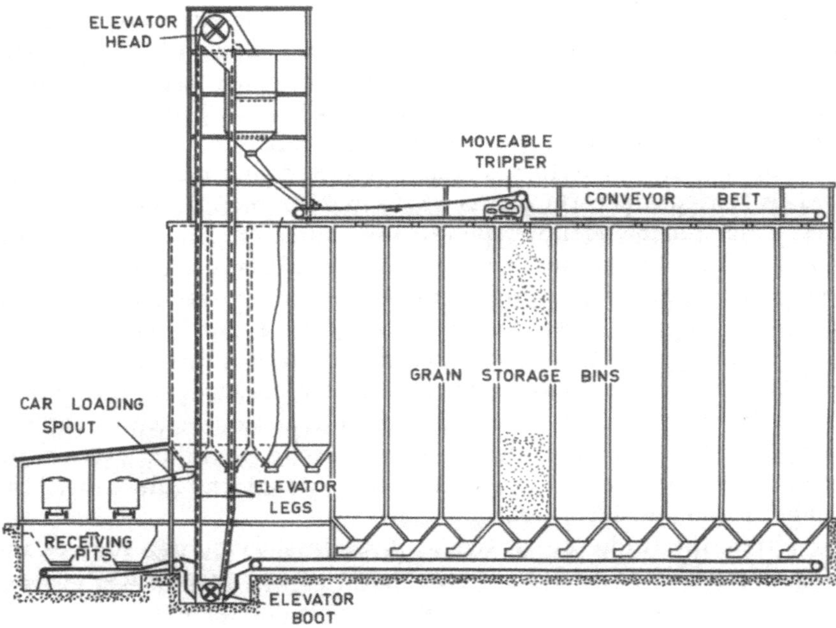

FIGURE 1.1. A typical grain export terminal.

The explosions were similar in several ways: (i) the premises were relatively new; (ii) the construction was reinforced concrete; (iii) both lacked adequate provision for explosion relief; and (iv) the office buildings were located near the head house.

Case Histories—Grain Elevators

Continental Grain Elevator, Westwego, Louisiana: Thirty-six Killed; Estimated Damage £15 Million

The facility consisted of 73 silos; five large metal grain tanks; train-unloading, barge-unloading, and ship-loading facilities; a head house which is a towerlike building containing grain conveyors, scales, and other equipment for moving, cleaning, and sampling grain; and, adjacent to the head house, an office building which, in addition to offices, housed a laboratory, lunch room, and control room.

The silos and head house were constructed of reinforced concrete. The silos were in three groups. They were all 35 m high but of three different diameters—7.6 m, 7.9 m, and 9.4 m. The gallery, basement, and an elevated conveyor structure above the silos had provision for explosion relief. Some of the silos were reported to have interbin venting. The top 23 m of the 76-m high head house was constructed of lightweight metal panels for explosion-relief purposes. Although some explosion-relief venting had been provided, it appeared to be inadequate.

The explosion occurred at about 9:05 am when some 75 people were on the site. It resulted in the top 40 m of the head house collapsing onto the office building and a series of explosions that continued throughout the morning in two of the groups of silos.

When area fire brigades reached the scene they found a mass of concrete, steel, and burning grain where the head house and office building had been. One man was rescued by helicopter from the top of one of the silos. The bodies of 35 fatalities were located in the rubble; 25 of them had been in the office building. One of the injured died several weeks afterwards.

In addition to the destruction of the head house and office building, the explosions and fires destroyed more than half of the silos; these were in the 1959 and 1962 groups. Many were blown out at the top and at the sides near the bottom. Few of the silos in the 1977 group had contained grain, but their conical metal bottoms had imploded and most of the silo tops of this group were also destroyed.

Farmers Export Grain Elevator, Galveston, Texas: Eighteen Killed Estimated Damage £12 Million

The facility came into service in the summer of 1976. It included (i) 60 reinforced concrete silos each 7.6 m in diameter and 38 m high with 178-mm walls and a 127-mm roof; (ii) two large metal grain tanks; (iii) a reinforced-concrete-built, 70-m high, head house; (iv) a single-storied office building of noncombustible construction; (v) a freight-wagon unloading shed built of metal on a metal framework; and (vi) a reinforced-concrete tunnel housing three conveyors connecting the unloading shed with the head house. (This tunnel ran under the office building.)

No effective explosion-relief venting had been provided for the head house, silos, or conveyor tunnel. Each elevator leg in the head house had an explosion-relief panel measuring 1.68 m × 1.29 m in the head section.

The freight-wagon unloading shed had facilities for bottom unloading three wagons at a time into the three pits provided. One pit could also receive grain by tipping but on the night of the explosion all three pits were using bottom unloading.

Prior to the explosion, the conveyor belt carrying grain to the northernmost row of silos had been stopped for repairs to be made. The motor driving the belt was in the basement of the head house. The employee who carried out the repairs had just radioed the control room to restart the belt when the explosion occurred, but it is not known whether the motor had been started.

The explosion in which the 18 people were killed and 22 injured occurred at 8:31 pm. It destroyed the freight-wagon unloading shed, the tunnel between this shed and the head house, and the central control room. The head house and office building were so badly damaged that they had to be demolished. Grain tank No. 1 collapsed and grain in the second tank caught fire.

A large ditch was created by the destruction of the tunnel, and this cut off the only road giving access to the site for fire fighting and also cut the main water distribution line.

Six of the fatalities were in the unloading-shed area, nine in the head house, two in the control room, and one outside. Fortunately, because of the late hour no one was in the office building.

Consideration of Both Incidents

In neither case was the source of ignition determined. The Galveston explosion is thought to have started in the area of pit No. 2 in the freight-wagon unloading shed and to have traveled through the tunnel into the head house. Possible sources of ignition included a diesel-electric locomotive that was over pit No. 2 at the time. Dust accumulations were reported to be a problem at the Galveston site.

Humidity levels were taken into account but, although low for these humid areas, they were not considered low for many areas where grain is handled.

The inadequate explosion-relief venting allowed the explosions to build up more pressure and cause more damage than would be expected if sufficient venting had been provided.

The incidents are not atypical of the problems that may occur at any time. A combination of the lack of explosion venting or other forms of protection and the layers of dust throughout the plant turned these incidents into major disasters, and the siting of office buildings near to an unvented dust-cloud area added considerably to the death toll.

It is not always realized by other industries that these two incidents

TABLE 1.1. Explosions, Deaths, Injuries, and Fires in Grain-Handling Facilities for 1958–1978

Year	Number of explosions in U.S.	Number of deaths	Number of injuries	Estimated number of fires in elevators[a] (NFPA)
1958	10	2	27	3200
1959	10	3	18	2200
1960	12	4	18	2300
1961	10	0	17	2100
1962	9	3	51	2300
1963	14	3	30	2200
1964	8	3	22	2000
1965	9	2	5	1900
1966	14	2	22	2000
1967	17	1	14	3000
1968	16	12	38	5300
1969	6	4	13	4700
1970	10	1	14	3000
1971	10	4	14	3100
1972	8	7	23	2400
1973	8	2	10	1800
1974	15	13	37	2200
1975	9	4	19	2200
1976	22	22	82	—
1977	21	65	84	—
1978	12	7	47	—
Totals	250	164	605	

[a] Estimated number of fires in elevators unavailable after 1975.

were statistically unusual only in the high loss of life. Dust explosions causing considerable damage have been a problem in the grain industry for many years with 8–15 major incidents occurring in the United States each year (Table 1.1). For this reason the grain industry has been the subject of the most extensive analysis of free-explosion statistics carried out for a single industry. Chiotti and his co-workers at Iowa University (1976) have looked at dust explosions in the American grain industry between 1958 and 1978.

Figures 1.2(a) and (b) summarize some of their results. It can be seen that in nearly 50% of the 162 accidents investigated the ignition source was not identified. The two most common sources found were in accidents associated with welding and faulty equipment that caused friction, such as the rubbing of conveyor belts.

In mills, foreign material in the grinding process was also a major source of fire and explosion. Further statistics on ignition sources published by the United States Department of Agriculture are given in Table 1.2, which lists 66 incidents between 1969 and 1978.

Of course it is dangerous to draw too many general conclusions

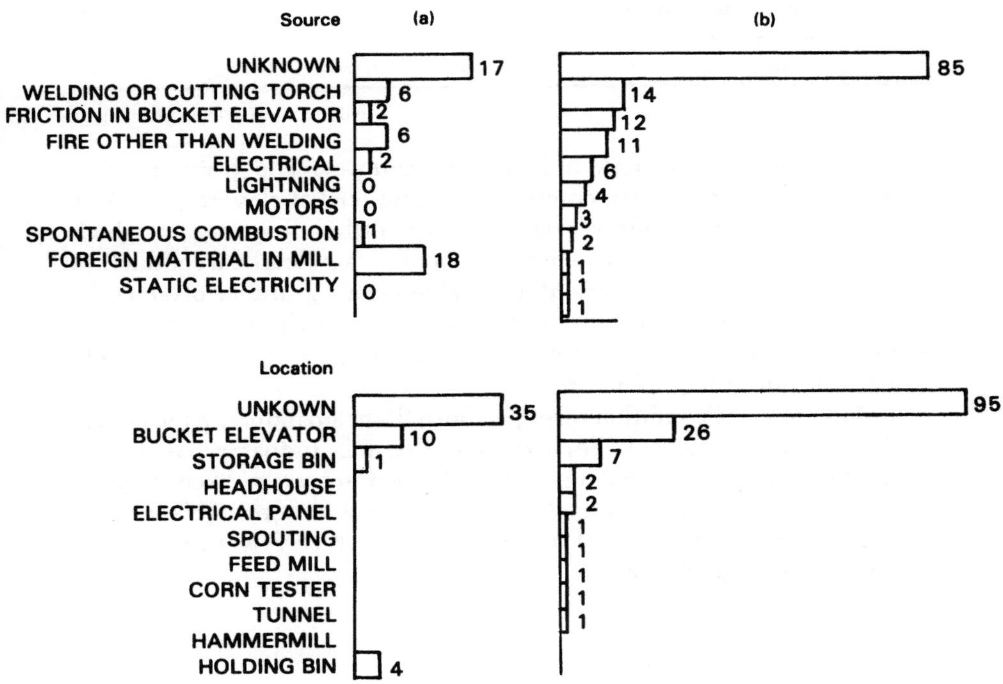

FIGURE 1.2. (a) Causes and locations of explosions in mills in U.S., 1958–1975 (50 incidents). (b) Causes and locations of elevator explosions 1958–1975.

TABLE 1.2. Probable Ignition Sources for 66 Incidents in Elevators

Ignition source	Number of elevators	Percent of elevators
Unknown	17	25.7
Welding/cutting	16	24.3
Hot bearings	7	10.6
Lightning	1	1.5
Static electricity	1	1.5
Electrical	4	6.0
Tramp metal	6	9.1
Blockage in leg	1	1.5
Extraction of oil from corncake	1	1.5
Switch engine on rail dump	1	1.5
Explosive vapor	2	3.1
Heating system	2	3.1
Dust system	2	3.1
Choked leg	1	1.5
Electric cord in leg	1	1.5
Volatile solvent escaped from processing of soybeans	1	1.5
Grain hung up in drier	1	1.5
Fan blade caused spark	1	1.5
Totals	66	100%

from an analysis of a single industry and country. The grain industry tends to transport by open conveyor belt and bucket elevator, and it is primarily a transport and storage industry rather than one in which processing predominates. The product shows a fair electrical conduction and has a high fine-dust content associated with a granular product. In the United States, the industry also suffers from the problem of old equipment that has been running many years and is not easily adaptable to modern safety requirements.

Many of these conditions are specific to the grain industry and do not apply to other industries. For example, the milk industry has a high percentage of drying problems, and polymers are more likely to give electrostatic sparks. However, it is only by a consideration of all possible problems that the hazards of an industry for which few statistics are available can be considered.

1.3 IGNITION SOURCES

Although opinions differ as to the relative importance of different ignition sources in different industries and countries, there is more or less universal agreement that the most common cause of accidents involving dust clouds is the use of welding torches or power tools without proper precautions and proceedures.

Welding torches or flames provide a very high energy that easily ignites any residual dust near the welding area. Even if maintenance is carried out only when no powder is transported, residual dust layers can ignite and smoulder and be swept into a dust cloud when the plant is restarted. A similar problem can be caused by cigarettes smoked by maintenance workers who think the plant must be safe when shut down.

Although high-energy sources of this type still cause the largest number of fires, other less-energetic sources must not be neglected.

For some time it was believed that dusts were much harder to ignite than vapors and that a major ignition source like a welding torch, electrical arc, or gross overheating was required before an explosion could occur. As time progressed and techniques for producing dispersed dust clouds in laboratory test equipment improved, the minimum energy capable of igniting a dust cloud was reduced. Now some 10% of organic materials ($< 75\ \mu m$ in diameter) can be ignited by a spark containing less than 5 mJ of energy, and some materials such as aluminum and sulfur have been ignited with sparks of less than 1 mJ. (Gases and vapors such as propane, methane, and many organic solvents have an ignition energy at their optimum concentrations between 0.15 and 1 mJ). Therefore, it is necessary to pay special attention to the low-energy sources of ignition such as electrostatic sparks and impact sparks from falling objects, if only because these often have not been considered in detail in the past because they were considered to be too low in energy to be of significance.

Electrostatic sparks, in particular, are often poorly understood in industry. A number of incidents occur each year which could easily be prevented if the importance of low-energy ignition sources were realized.

Although the vast majority of accidents continue to be caused by plant failure, human error, and failure to observe specified rules, it is important that low-energy ignition sources should not be neglected in favor of the more obvious high-energy sources.

In this book special attention is given to low-energy sources in general and electrostatic problems in particular. Electrostatics remains a poorly understood subject in industry and static sparks become the scapegoat when no other source of ignition seems likely. The aura of mystery that often seems to surround the question of electrostatics is not justified.

Static levels can be quantified and hazards evaluated, and in Chapters 3 and 4 the means by which this is accomplished are specified. Impact sparks require additional research because laboratory tests have not yet succeeded in igniting an organic dust cloud by an impact spark unless a chemical process such as the thermite reaction is involved. However, the statistics of accidents still quote impact sparks as a probable cause of ignition in a significant number of industrial dust explosions each year. The current state of knowledge on ignition by impact and friction is reviewed in Chapter 2.

Standard tests on dust ignitability generally lead to two minimum ignition temperatures: the temperature of a surface on which a dust layer must be placed in order to ignite and the minimum temperature of an oven which will cause a dust cloud to ignite.

A number of accidents have occurred which show that these tests are inadequate. With thick layers or heated heaps of powder, a slow self-heating can occur because the heat created by oxidation or decomposition is unable to escape due to the poor thermal conductivity of a powder. More extensive tests have recently been developed to detect this problem and the application of the theory of thermal runaway to dust deposits is now much better understood.

Thermal ignition sources are discussed in Chapter 2 and the different test methods now used are described in Chapter 5.

1.4. IGNITION PROPERTIES OF POWDERED MATERIALS

Flammable gases may be easily classified by their *ignitability* because the mixture of gas and air is uniquely defined once its concentration is known. Ignition of a powder is the result of a chemical reaction of the surface of the particles. The surface to volume ratio, and therefore the particle-size distribution, considerably affects both the ignitability of the dust and the results of the explosion. As the powder becomes finer the dust becomes easier to ignite, and the pressure rise and rate of pressure rise during the explosion increase. Different samples of the same material are likely to give different results in explosion tests if the size distribution differs. Tests have shown that the explosive properties do not vary much if the particle size is less than ~50–75 μm, but the ignitability and the results of the explosion become less severe above this size. Ignition tests are normally carried out on particles that have passed through a 75-μm sieve.

As a matter of convenience the word "dust" is often used when referring to the fine fraction and the word "powder" to a product that consists mostly of particles above this limit.

A range of standard tests have been developed to define the ignition characteristics of powdered materials and these are described in Chapter 5. The ignition of a powder or dust is complicated and different laboratory test systems tend to give different results. Ignition is affected by size, shape, agglomeration tendency, and minor surface differences, as well as by effects of the apparatus such as volume and turbulence. Therefore, dust ignition tests cannot be treated as precise measurements in the way that gas ignition characteristics can, but they do provide a useful guide to the relative behavior of different materials.

In general, it is necessary to consider very carefully the conditions that prevail in the factory. This must be done to ensure that laboratory tests fairly represent the true situation and also to ensure that the test sample is in the same state as the product handled, with the same particle-size distribution and contaminants.

1.5. DUST-CLASSIFICATION SYSTEMS

In general terms dusts are classified according to the ease with which they ignite or the results of any ignition. British sources have a very broad classification. Group-A dusts will propagate a flame in the Fire Research Station test apparatus while group-B dusts will not. Three types of test apparatus are used, a vertical tube in which dust is dispersed upwards through an electric spark, a horizontal tube with a hot-wire ignition source at 1300°C, and an inflammator in which dust is dispersed downwards past a source which may be either an electric spark or a heated wire. These tests are described in detail in Chapter 5.

Although group-B dusts do not explode in the test apparatus, they may ignite with extremely high-energy sources or at elevated temperatures. High air pressures or the presence of a small quantity of flammable vapor may also render the dust explosive.

In the United States Fire Codes, Class I Hazardous Materials refer to flammable gases and vapors, subdivided into classes A–D. Class II hazardous materials are dusts with subgroups E, F, and G. For example, group-E atmospheres contain metal dust, group-F atmospheres contain carbonaceous dusts, group-G atmospheres contain flour. Class III hazardous materials refer to fibres and filings.

A hazardous location in the United States is given a class and subgroup defining the type of flammable material present, and a division to define the probability of finding the material present at any time. Generally, Division I implies that the flammable atmosphere or layer will be present under normal operating conditions, or where faulty operation might cause the flammable atmosphere or a layer to occur.

Division II implies that such flammable conditions are unlikely. A full definition may be found in the *United States National Electrical Code*, published by the National Fire Protection Association.

In 1962, the U.S. Bureau of Mines described powders by an index of explosibility in which explosion parameters were compared to a standard dust taken to be Pittsburgh coal (Jacobsen *et al.*, 1962). The index of explosibility was defined as the ignition sensitivity multiplied by explosion severity. Ignition sensitivity was the ease with which a dust could be made to ignite, and ignition severity measured the results of ignition defined in terms of the maximum pressure produced and the rate at which it rose. Powders were compared to a standard dust from Pittsburgh coal. The explosibility is given by the ratio of the products of the ignition characteristics: that is,

$$\frac{\text{Minimum ignition energy} \times \text{minimum ignition temperature} \times \text{minimum explosion concentration (Pittsburgh coal)}}{\text{Minimum ignition energy} \times \text{minimum ignition temperature} \times \text{minimum explosion concentration (sample dust)}}$$

and the explosion severity by minimum ignition energy:

$$\frac{\text{maximum explosion pressure} \times \text{maximum rate of pressure rise (sample dust)}}{\text{maximum explosion pressure} \times \text{maximum rate of pressure rise (Pittsburgh coal)}}$$

This system, by presenting explosibility and explosion severity in comparison to coal dust removes some of the uncertainties introduced by different test equipment. However, it requires a full range of tests to be carried out in order to characterize each material and is a rather complicated procedure.

In Germany and Switzerland dusts are characterized by the rate of pressure rise measured in an ignition in a 1-m^3 closed vessel. This is the most relevent parameter in designing explosion-protection systems that will be installed in areas where an explosive dust cloud might occur. Ignition energy and temperature are of less importance, except as a measure of the likelihood of ignition.

In fact, it appears that a measure of ignitability can also be obtained from the graph of the rise in pressure as a function of time during an explosion. The induction time between a spark source of ignition and the time at which the pressure starts to rise has been shown to be proportional to the minimum ignition energy in a large number of cases (Bartknecht, 1980).

Dusts are divided into classes 1, 2, and 3 generally known as St. 1, St. 2, and St. 3, where St. stands for Staub, the German word for dust. Dusts in St. 1 have a maximum pressure rise rate of <200 bars/s in a 1 m^3 chamber. This group includes the majority of common dusts. St. 2 dusts have pressure rise rates between 200 and 300 bars/s, while St. 3 dusts have pressure rise rates exceeding 300 bar/s.

In all cases, the ignitability and explosion severity are a strong function of particle-size distribution. In most standards only particles passing through a 75-μm sieve are tested, but differences are also observed as the distribution within this range changes.

Variations occur not only with particle size and shape but also with changes in test apparatus, such as size and shape of the test chamber, and the degree of turbulence of the dust cloud. Thus, when published figures are used to design equipment, explosion parameters must be used to provide a general guide rather than precise design criteria. For instance, it is possible to use standard tests to identify whether low-energy ignition sources, such as electrostatic sources, must be considered or whether the energy required for ignition is above the likely limit of this type of source without being very specific about the exact energy required for ignition. It is also possible to classify dusts to enable safe venting conditions to be chosen, but it is not as yet possible to find unique values for standard explosion parameters measured by different authorities. A variation by a factor of 2–3 is not unusual.

1.6. PROTECTION SYSTEMS

Since a good safety margin is always allowed, precise measurements of ignition parameters are not required for the design of the safety precautions which must be taken in case an ignition occurs.

The four ways in which a plant may be protected from an ignition are as follows:

(i) Venting—a weak section or trap door is placed in a suitable position so that excess pressure in an explosion can escape rapidly without causing damage.

(ii) Suppression—the initial stages of the explosion are detected and initiate the release of an extinguishing material.

(iii) Inerting—the oxygen level in the equipment is reduced to a level that is too low to allow oxidation to proceed rapidly enough for explosion.

(iv) Containment—the equipment is built strong enough to contain the explosion and thus eliminate any external damage.

If an explosion occurs three conditions must occur simultaneously: (i) there must be sufficient oxygen to support combustion; (ii) there must be an ignition source of sufficient energy; and (iii) there must be a dust cloud within the explosive density limits.

It must always be considered impossible to be totally certain of eliminating all ignition sources unless the powder concerned is extremely difficult to ignite. Therefore, if an explosive dust cloud could occur, under any circumstances, some form of protection that prevents plant damage and loss must be applied. The four protection methods described above essentially remove one of the requirements for ignition or allow the explosion to occur in a controlled manner.

The choice of protection methods depends on the size of the plant, the type of product concerned, and the economics of the situation. The factors that influence this choice and the details of the different methods of protection are described in Chapter 7.

1.7. GENERAL CONSIDERATIONS

Although some fires and explosions occur because of unforeseen factors in the product or equipment, the vast majority are still due to human error or a failure of equipment or procedures.

Some of the most important elements in avoiding fires and explosions and minimizing the damage caused when an accident occurs are education of the work force, effective testing and maintenance, detailed and foolproof procedures of operation and shutdown, and good housekeeping. The importance of these items is clearly shown in many of the case histories described throughout the book. The safety of operations must be considered in detail throughout every stage from the initial conception of the plant to its completion and throughout its operating life.

2

Ignition Sources

2.1. WELDING, CUTTING, AND FLAMES

The statistics of fire and explosions in nearly all industries and countries still attribute the majority of accidents not due to malicious damage to heat generated by flames, welding torches, or cutting tools. These operations generate many times more energy than is required to ignite dusts, either in cloud or layer form.

The accidents occur in one of two ways: (i) A procedural error may occur. For example, if communications are poor, equipment under maintenance may be started in error or it may be put into operation in mistake for another piece of equipment. This is most likely to occur when there is a centralized control room or control point. (ii) Insufficient cleaning of the maintenance area can result in dust catching fire some distance from the welding point.

Dust layers often ignite at temperatures of only 100–200°C; therefore, deposits of powder must be removed from all surfaces that may become hot during welding. It is not sufficient merely to clean the plant area. Dust layers *inside* the equipment must be cleaned out because if residual dust overheats, smoldering layers of powder can remain inside the equipment for several days until the plant is restarted.

Another common problem during maintenance is cigarette smoking. The no smoking regulations are often ignored during maintenance periods, partly because it seems illogical to ban a small ignition source when major heat sources such as welding torches are present. The workers often feel that if the plant is not operating the danger is not present. Careless discarding of cigarette ends is an important source of fires and explosions. If there are any deposits of dust outside the plant they can ignite and burn very slowly, remaining undetected for many days.

Flames are another obvious danger in processes where waste

products are burned. In any such process, the heating that may occur in all fault conditions must be considered carefully. Accidental fires may occur in any factory and these can form an ignition source for dust clouds, resulting in an explosion. Flammable materials such as packaging should not be stored close to regions where there is a dust cloud.

Powders whose size is above the explosive limit may burn if ignited by a high-energy source and can then act as an ignition source for a fine dust in another part of the plant. Housekeeping procedures should include removal of all flammable materials from the vicinity of high-energy sources such as welding torches and flames.

The avoidance of fires and explosions caused by the high-energy sources from flames, fires, and welding is primarily achieved by strictly enforced procedures, careful housekeeping to eliminate dusts or powders in any form from areas where any of these sources might occur, and effective fire-prevention systems that detect and extinguish any minor fires immediately.

It is very easy for a dust deposit to continue to smolder for many days with little external sign, and it is very difficult to extinguish a fire in a dust deposit because the particles have a good air supply. Therefore, any extinguishing material used must penetrate the dust pile without disturbing the pile and forming a dust cloud. Small fires may form an ignition source long after they are believed to be extinguished.

Case Histories—Welding and Flames

Grain Storage System: U.S.

Welding equipment was in use at the top of a silo when bucket elevators were started in error. The resulting dust cloud ignited and there was one fatility.

Sawmill: U.S. 1978

Approximately 60 sprinklers held the loss to $15,000 in a flash fire at a sawmill. The incident apparently involved escaped acetylene gas that had collected beneath the roof.

On the day of the fire, the mill was shut down for installation of a new slabber chipper and a new infeed chain for the debarker. The installation of the latter involved flame cutting on the outside of the conveyor. At the same time, welding was being done in the space below the debarker.

A 11:00 am, what was described as a "low-noise explosion" occurred in the debarker area. This quickly activated approximately 60 heads of the overheads dry-pipe sprinkler system. The installation crew turned off the acetylene and oxygen cylinders and called the fire department. Damage was mainly to the wood debarker-operator's shed, wiring for the debarker controls, and two control boxes.

It is believed that a welding spark hit the rubber hose that supplied acetylene to the welders. This could have caused the hose to burn through, releasing a large quantity of the gas which, being lighter than air, would have collected immediately beneath the roof. There was no sawdust accumulation at the ceiling that could have caused the flash fire.

LUMBERYARD, U.S. 1975

A fire in the ductwork feeding a metal sawdust collector made it necessary to empty the sawdust collector. The ductwork was disconnected and the fire in it was extinguished by plant personnel. Early on the following morning, the Fire Department received an alarm reporting another fire inside the sawdust collector.

Although the collector had been emptied, several large deposits of sawdust were caked in its corners and crevices. At least one of these deposits was smoldering. Fire fighters pulled two booster lines off their engine and placed a ladder inside the sawdust collector. A fire fighter climbed inside the dust collector and loosened a small amount of the caked sawdust. As the smoldering sawdust fell, it began to burn. Before water could be directed onto it, a small explosion occurred. This loosened additional sawdust deposits, which quickly ignited. Fire fighters directed the water spray into the collector and extinguished the fire.

The fire fighter who was inside the collector and two other fire fighters who were within 12 ft of it were burned. The fire fighter in the dust collector died from his injuries 4 days later.

2.2. SELF-HEATING—SPONTANEOUS COMBUSTION

Spontaneous heating can occur whenever there is an exothermic reaction that generates heat faster than it can escape. Since the rate of heat generation often increases as the temperature rises, in the worst case, a runaway condition occurs and the reaction proceeds to ignition.

Oxidation is the most common reaction involved. Fats and oils react readily with oxygen, and some finely divided metals are also prone to self-heating particularly if they are produced in an inert atmosphere to prevent dust explosions. The surface is then unoxidized and highly reactive on exposure to air.

The action of microorganisms can bring about significant temperature increases. Although microorganisms are usually killed at about 70° C, well below ignition, the heat produced can be sufficient to start a chemical reaction, such as oxidation, which then proceeds to ignition. Since water is necessary for microbiological activity, closely stacked damp products are particulary susceptible.

Other chemical reactions can also produce heat; for example, some materials react exothermically with water. Quick lime, although not itself combustible, can produce sufficient heat to ignite surrounding combustible materials. Acids often produce heat in reaction with wood or chemicals, and polymerization or decomposition may also be exothermic. The rate at which most chemical reactions proceed generally follows the Arrhenius equation

$$K_T = A \exp(-W/RT)$$

where K is the rate of chemical reaction, T the temperature, R the universal gas constant, W the activation energy of the chemical reaction concerned, and A a constant.

Although in complex chemical reactions it is difficult to relate W and A directly to a parameter with simple chemical significance, the rule that the reaction rate increases with temperature following an exponential law is general.

An exothermic reaction, therefore, generates heat that increases the reaction rate which gives more heat, which increases the reaction rate still further, until the reactants are used up. As the temperature rises, heat loss by conduction, convection, and radiation also increases and a steady state may be reached where the heat loss balances the heat produced. A reaction may be slowed or stopped by a number of effects: for example, if one of the reactants is used up, if a material melts or evaporates (so a considerable amount of latent heat is absorbed), or if a passive film is formed preventing contact between the reactants.

Under these conditions, the temperature may level off even though the direct balance between heat production and loss has exceeded the initial point where the temperature starts to increase.

2.2.1. Self-Accelerating Reactions

Most reaction rates increase with increasing temperature but some reactions, known as *self-accelerating reactions*, also speed up if the temperature remains constant because of changes that occur in the reacting materials. For example, in autocatalysis the reaction rate increases by the action of one of the products of the reaction itself. Alternatively, an inhibitor may be blocked or consumed by the reaction of one of its products. A change of phase or degree of aggregation of particles can also increase reaction rates. Self-acceleration reactions may not always be obvious in routine thermal testing because there may be a long time delay before the reaction occurs. A product should be regarded with suspicion if routine tests show a faster-than-normal acceleration of temperature or a high activation energy.

If it is known that other similar products or classes of reaction are self-accelerating, then special long-term thermal tests should be carried out to look for anomalies.

2.2.2. The Theory of Self-Heating

As the temperature of a reacting material increases, the rate of heat production increases exponentially. If heat loss is primarily by conduction (as would be the case in a heap of powder), then the heat loss is linear with temperature. Therefore, an ambient temperature exists, known as the *critical temperature* T_c, above which equilibrium is not possible. This temperature is defined exactly but depends on all the parameters that affect heat loss, such as the size of the deposit, the air flow through or across the powder, the thermal conductivity and packing, and the temperature gradient through the deposit.

The accumulation of heat in the system requires a finite time which depends on the ambient temperature and volume of powder concerned. This is known as the *induction time* and is an important characteristic of spontaneous-ignition reactions.

The earliest theoretical treatment of spontaneous combustion was made by Semenov in the late 1920s and early 1930s (Semenov, 1928, 1935). He made the assumption that the reactant was uniform in temperature, which is only true of a stirred liquid, or in rare cases, a powder stored at a high temperature. A more realistic theory, applicable to dust deposits where heat transfer occurs through the layer primarily by conduction, was developed by Frank Kamenetski (1939). Both theories have been developed over the years and extended so that the effect of the consumption of the reactants can be included. The theories are reviewed by Tripper (1967).

Experimental laboratory tests enable the reaction rates and activation energies of a reaction to be determined. The thermal conductivity of a dust deposit can also be measured and, with this data and a knowledge of the temperature profile and geometry of dust deposits in the plant, the critical ignition temperature of a dust deposit, where heat production exceeds heat loss, can be predicted.

Examples of comparisons between theoretical and experimental estimates of ignition temperatures of specific products are given by Bowes (1968) and Merzhanov (1967). Both authors stress the importance of ensuring that all reactions are included in the analysis—both those that initiate self-heating and those that occur only once self-heating has begun. In many processes ignition occurs in the vapor phase, so vaporization must also be brought into the calculation. When all these considerations are taken into account, it may be possible to extrapolate results obtained from laboratory samples to industrial conditions.

Leuschke (1980) has investigated the effect of the volume of a deposit on its ignition temperature. In these tests, the dust was suspended in an oven in a wire-mesh cylinder whose length was equal to its diameter. The ignition temperature was defined as the ambient oven temperature at which ignition occurred in a test where the oven was held at constant temperature until self-heating itself increased the oven temperature.

In these tests, the self-ignition temperature decreased with increasing volume according to a logarithmic law

$$\frac{1}{T_c} \propto \log V$$

It was also found that the induction time at the lowest ignition temperature increased with volume. The logarithm of the induction time was proportional to the volume, with the times involved differing widely from material to material. Results from a starch product and coal are given in Figure 2.1.

The induction time decreases faster the higher the temperature is above the critical temperature at which self-ignition first occurs, and at higher temperatures the dependence on volume becomes progressively less logarithmic, tending to become independent of volume. In these tests, the induction time included the time during which the sample was heating up to the ambient oven temperature. This is obviously longer for larger-thermal-capacity samples. The results do not apply to stored-hot-dust situations, in which case increasing the sample size simply reduces heat loss from the center with a resulting decrease in the time to ignition.

The complicated nature of self-heating reactions in practice is illustrated in the U.S. Bureau of Mines report on the spontaneous heating of coal (1977). The rate of spontaneous heating of coal is affected

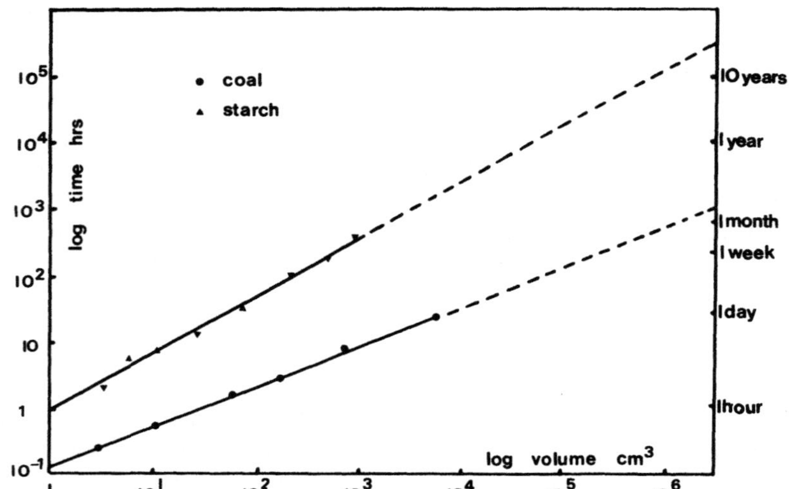

FIGURE 2.1. Induction time at lowest-measured self-ignition temperature as a function of dust volume.

by air-flow rate, particle size (small particles self-heat faster), rank, moisture content, temperature, and pyrite content. Of these, rank (i.e., coal composition) and moisture content have the greatest effect. Below 100°C the heat of wetting is greater than the heat of oxidation. The gaining of moisture is an exothermic process, therefore the passage of moist air over dry coal causes it to heat up.

2.2.3. Prevention of Self-Heating

The temperature at which ignition occurs depends on the cooling of the deposit and on its volume. Therefore, the results of tests to identify the ignition temperature depend on the apparatus chosen. The worst case occurs when there is no cooling of the dust deposit, i.e., adiabatic conditions. If tests indicate that at the optimum temperature for the process and adiabatic conditions (no heat loss) the reaction cannot proceed to ignition, then problems only exist if there is a fault that causes abnormally high temperatures in the system.

However, some processes require temperatures to rise above this safe limit. In these cases special precautions must be taken both to increase cooling and to ensure that the behavior of the product under different cooling conditions is thoroughly known and understood. Changing from batch to continuous processing may decrease the hazard because less material is stored and cooling is easier. Automatic monitoring of temperatures to detect any temperature rise above the norm may also be useful if the location at which overheating may occur is well defined. This is likely to be of use in storage systems but less practical in driers or extraction systems where surfaces may overheat unintentionally. The hazard is also reduced if the time for which a powder remains at an elevated temperature is reduced.

In some cases it may be possible to change the process so that a reactive component, or a material which acts as a catalyst, is added at a later stage and does not have to be heated.

Elimination of self-heating requires consideration not only of known hot areas such as driers, but also of surfaces which may become overheated if a fault that increases heat production or prevents cooling occurs. Many processes that do not intentionally heat a powder will, in fact, have this effect. For example, powders from grinding machines are frequently warm. The ambient temperature of a sensitive powder should be known throughout the system. At fairly low ambient temperatures the temperature rise due to self-heating may be extremely slow and may have remained undetected in laboratory tests. In this case a very long induction time may be needed before the temperature rise becomes significant. It is therefore advisable to avoid dust deposits that may remain undisturbed for long periods of time. This tends to occur in crevices, pipe bends, or on horizontal surfaces, unless care is taken in design to ensure that this cannot occur.

Fires and explosions due to self-heating often occur at temperatures below that at which a problem is believed to exist. In some cases this is due to a change in the product itself; for example, in its composition or size distribution. But there is also a considerable problem in interpreting published figures for minimum ignition temperatures. The methods by which ignition temperatures are measured are described in Chapter 5 and it can be seen that techniques used by different workers produce widely different results. It is important to match the test conditions to those prevailing in the plant and to ensure that the powder tested is in the same state as in the factory. For example, a newly produced product may undergo a slow reaction at ambient temperatures and be in a different form at the time of the tests than when it was freshly produced.

2.3. HOT SURFACES

Hot surfaces provide a source of ignition to dust layers resting on them. Although these layers tend to burn rather than explode, if the layers are disturbed the concentration may exceed the maximum explosive limit and an explosion can occur. Many of the considerations that must be applied to layers of dust were described in detail in the previous section, because ignition of powder layers generally involves a self-heating process.

The minimum temperature of a hot surface needed for ignition of a dust layer in contact with it decreases as the depth of powder on it increases. The thickness of the layer is the most important influence on the ignition temperature. Particle size has very little effect and packing of the layer is important only for thin layers of less than 0.5 cm. The thermal analysis developed for self-heating in bulk products can be applied to layers on heated surfaces. The heat loss is both by conduction through the sample and convection above it. This calculation has been carried out by Bowes and Townshend (1962).

Once again it is necessary to be sure that all possible reactions are included in the analysis of the total heat produced, including the reactions that first initiate self-heating and subsequent oxidation and decomposition reactions.

It has been shown that the dependence of the ignition temperature on dust thickness can be described by the equation

$$\log (\gamma_c T_c^2/r^2) = (M - N)/T_c$$

where r is the half-thickness of the layer in millimeters, γ_c is a dimensionless parameter (normally 0.88 for dust layers), and T_c is the ignition temperature. M and N are the only unknown constants; therefore two measurments of the ignition temperature T_c and the layer half-thickness r allows M and N to be defined. The ignition temperature of any layer thickness can then be estimated. The minimum ignition temperature may be measured using a number of different geometries, layer or bulk, and the temperature obtained is a function of the geometry chosen. Measurements made in one geometry cannot be applied directly to another geometry unless due allowance is made.

The behavior of a heated powder may differ according to the way it is heated. Bowes and Townshend (1962) found that a 0.3-cm layer of sawdust could not be ignited at temperatures up to 410°C while a 0.5-cm layer ignited at around 320°C. However, if the thin layer was deposited rapidly on a preheated plate the ignition temperature was more comparable with those for thicker layers. The ignition temperature is also likely to be lower if the ambient temperature is high since heat loss from the layer will be reduced.

Hot surfaces may be produced in any process that directly involves heating such as drying, but they may also occur inadvertantly in other circumstances, such as from lighting or machinery or in grinding processes where a lot of mechanical work is done.

Powder layers are very poor thermal conductors so deposits of powder may cause surfaces to overheat by preventing normal cooling. The surfaces of electrical motors, machinery, and electronic equipment must be kept clean of dust layers that inhibit cooling.

Overheating can also be caused by friction due to the ingress of dust into moving parts. Therefore motors and bearings should always be dust tight. Since friction in any form will create heat, slipping conveyors may overheat to such an extent that the belt or the product it is carrying will ignite. If dust can collect between the slipping surfaces it will be heated directly by the friction of the belt and roller surfaces rubbing against the particles. Because dust is a poor thermal conductor, it may become considerably hotter than the macroscopic temperature of the metal roller against which it is rubbing. Therefore the roller temperature is not necessarily a good guide to dust temperatures between slipping surfaces.

Reports of fires and explosions often show that when a fire is attributed to a hot surface, the temperature of the surface is believed to have been well below the dust ignition temperature. In some cases, the surface has been found to be hotter than was believed, but in many cases the layer was found to be more ignitable than previously expected.

The importance of ensuring that laboratory tests adequately simulate

conditions which prevail in a factory has already been stressed but often the layers that build up in a plant are not the same composition as the pure product. For example, oil contamination from a small oil leak may reduce the ignition temperature of a layer or volatile materials from the product may collect in a layer and make it more easily ignited. Additional factors like these must be taken into account when evaluating a plant's safety.

Case Histories—Overheating

Cattle Food and Soap Manufacturer: U.K. 1972

Men returning from a tea break smelled something burning and saw flames coming from the base of one of eight 100-ton silos which had its base on the eighth story and extended to the tenth story. The silo was being filled with meal by a worm screw, thus creating a dust cloud within the silo. The mens' immediate reaction was to turn off the silo light since it was a common occurrence for meal dust to ignite on contact with a lit silo light bulb. However, before this could be done, the dust cloud in the silo ignited and an explosion occurred. The light bulb was situated in a well, but the thick protective glass was missing. Dust therefore came into contact with the hot light bulb. The fire subsequently spread over four storys of the 11 story plus basement building and further explosions occurred. The loss was estimated at £65,000.

Flour Millers: U.K. 1979

Heat generated by a slipping conveyor belt ignited the rubber belting this in turn ignited the dust. The total damage was £510,000.

Chemical Company: U.K. 1976

A poultry-feed additive was left in a closed drying vessel for 27 h after drying was complete. The thermal insulation in the dryer kept the temperature at 120–130°C. Under these conditions the product self-heated and there was a fierce explosion killing one man and doing £2 million worth of damage.

Insecticide Manufacturing Plant: Canada 1975

This plant mixed insecticides for horticultural applications. The batch mixing was done in a stainless-steel drum 5 ft long and 4 ft in diameter. The drum was loaded with various chemicals to a gross weight of 750 lb per batch. The drum was revolved for a minimum of 30 min to thoroughly mix its powdered contents.

Approximately 15 min before the end of the last shift on the evening before the fire, the mixing drum was loaded with expanded mica, dimethyl-dichloroethyl phosphate, wood flour, powdered milk sugar, ground rock, and potassium chlorate. The mixer was turned on for a maximum of 5 min, after which it was shut down as employees began to clean up before leaving the plant at the end of their shift. The drum was left loaded so that mixing could be completed the following day. Although this procedure was not common practice, it had been done several times in the past without incident.

Since potassium chlorate is a very strong oxidizing compound, an exothermic reaction can occur whenever it comes into contact with a combustible material, and an explosion could occur. Investigating authorities believe that the potassium chlorate oxidized the powdered milk sugar, causing an explosion in the mixing drum. They surmise that this explosion suspended the wood flour and remaining powdered milk sugar (originally in the drum) in the air. Furthermore, since there was always some accumulation of various dusts on the structural components and machinery in the mixing room, the authorities feel that the initial explosion would also have suspended this dust in the air. A second explosion is believed to have occurred in this suspended dust, causing far greater damage than that created by the initial explosion.

Loss was estimated at approximately $55,000.

Paper Manufacturers: U.S. 1975

Spontaneous ignition is believed to have caused this fire that burned in a huge wood-chip pile for over 2 weeks. The wood chips were used to make paper in adjacent plant buildings. The pile, which covered 5 acres and ranged in height from ground level to 70 ft, contained about 60,000 cords of wood chips. The fire also ignited several boxcars located on a nearby railroad siding.

The fire had been smoldering for several weeks when it was whipped out of control by high winds on November 10. At this time, the fire was beyond the capacity of the local Fire Department and threatened to spread to nearby homes and businesses. Equipment from three other towns and the Michigan Department of National Resources responded, along with additional help from local plants and a nearby Air Force base. Seven bulldozers were utilized to move the burning chips in order to expose fires below the pile's surface and allow better penetration of water from the hose streams.

As the wind velocity diminished over a period of several days, the blaze was brought under control to a point where only deep, smoldering fires remained.

The "fire emergency" condition was terminated on November 26, but the Fire Marshal ordered the paper company to maintain a full-time firewatch.

Over 300 fire fighters spent a total of 6674 work-hours combatting the fire, and over 43,774,000 gallons of water were pumped on the fire. Over 55 Fire Departments combined with several police organizations and the Air Force to contribute their manpower and equipment.

An estimated $2 million worth of wood chips were destroyed.

Grain Elevator: Canada 1975

A 160-ft-high grain elevator was operating when a fire broke out in a metal "leg" that contained a conveyor. The wood-framed, wood-walled elevator with a metal-clad exterior could hold 15 million bushels of grain. The workhouse, warehouse, and part of the gallery had sprinklers. When the fire occurred, the elevator was loaded with bran pellets which are known to cause dust problems.

On the morning of the fire, the elevator was unloading the bran pellets into a ship when a "garner" (an intermediate storage bin in the leg) became loaded beyond capacity. A safety overload switch is supposed to shut off the conveyor if the garner is full. The switch did not function, and the conveyor continued to bring more bran pellets into the garner. The garner overflowed and the excess bran pellets ran down the leg. They jammed the conveyor belt, which slipped on the still turning pully. Friction heated the belt until it ignited. The burning belt parted at the head of the leg and fell down the leg, igniting dust there. Employees tried to fight the fire with portable extinguishers but could not put it out. A sprinkler in the leg then functioned and seemed to control the fire, but it stopped working after about 30 s and the employees left the area. Investigators later found that other employees, who had heard the sprinkler bell at the ground level, assumed that the system was malfunctioning and had shut it off.

About this time, someone called the Fire Department. The nearest station was about a 3-min drive from the elevator. While fire fighters were responding, an explosion occurred, followed almost immediately by a second, more devastating blast. A large black smoke cloud was visible from a distance. When fire fighters arrived, flames were visible in the workhouse. Injured employees and debris were scattered about the area. Five people had been killed and 13 injured. The explosion also broke a water main under the elevator.

The fire chief ordered the ship moved away. He assembled two fire boats, two engines, two ladder trucks, an elevating platform, and 25 men to control the fire. Control was achieved in about 2½ h, but hot spots continued to burn for 2 days. Damage extended to nearby automobiles, railways cars, tracks, and the wharf itself. Dollar loss will probably exceed $8 million.

A coroner's jury investigating the fire made the following recommendations (i) maintain better housekeeping to minimize static dust; (ii) mark escape routes for use in case of fire; (iii) install better safeguards on the legs to shut down the system in the event of jamming; (iv) hold fire drills periodically; (v) have fire alarms with distinctive sounds for different areas; (vi) send a fireboat from nearby Vancouver to every waterfront area fire on the first alarm.

2.4. SPARKS

Sparks may be divided into three distinct phenomena (i) electrical sparks, (ii) electrostatic sparks, and (iii) frictional impact sparks.

An electrical spark occurs when a circuit fed by a voltage source is broken. The voltage across the spark gap is fixed by the voltage source. Often the discharge is initiated by very small gaps as the circuit is broken, but once the spark has begun the high current capacity of the source allows it to continue for a short time as the gap increases. An electrical spark may therefore be produced by a relatively low voltage source such as a 12-V car battery or a 220- or 110-V main power supply.

An electrostatic spark occurs when there is no external voltage source but a charge, created for example by friction, accumulates on an isolated conductor. The accumulated charge may raise the potential to several kilovolts even though the total charge stored is not high. A spark occurs when the voltage rises to a level sufficient to break down the air in the spark gap. Since the charge available is low, the current that can flow across the gap is also limited and the spark is of short duration. Electrostatic sparks are therefore a high-voltage phenomena in which a relatively low charge transfer, compared with an electrical spark, takes place. Electrostatic sparks are discussed in detail in Chapters 3 and 4.

The energy dissipated by a spark passing through air, whether the spark is electrical or electrostatic, is defined as the product of current and voltage throughout the duration of the spark, i.e.,

$$W = \int VI\, dt$$

In this case the voltage is the potential difference across the spark gap (not necessarily equal to the voltage source) and I the charge flowing in each unit of time (current).

The ignitability of a dust is often defined in terms of the minimum ignition energy. The test is usually performed with an electrostatic spark and it is unlikely to be precisely applicable to electrical sparks. However, electrical sparks are often significantly above the minimum energy that may be taken as a rough guide to a minimum safety requirement.

Impact sparks are not really sparks at all, but very small hot spots created when the energy of impact is dissipated as heat at the point of contact. It is not possible to define an impact energy in terms of current and voltage in the same way as for an electrical spark. The energy of an impact spark is most often defined by finding a gas mixture which is ignited by the impact, then finding the equivalent electrical spark that will ignite the same mixture. This method has a number of problems when applied to dust clouds which are discussed further in Section 2.4.3.

2.4.1. Electrical Sparks

When a circuit carrying an electric current is broken or there is a small gap in a circuit, the current may be able to flow across the gap and form a transient arc. The power dissipated in the arc or "break spark" is given by I^2R where I is the current flowing in the circuit and R the electrical resistance of the gap. If the gap resistance is 50 Ω and the current 5 A then the energy dissipated in the gap is 1250 J/s.

A short spark of only a few millijoules is sufficient to ignite many dust clouds and a major electrical arc of this type can ignite not only dust clouds and layers but also other flammable materials which may later ignite a dust cloud.

Some equipment, such as switches and make and break relays, necessarily give small sparks in normal operation. These must therefore be protected by being made completely dust tight or kept away from any area where dust may accumulate. Sparks may also be produced if a current-carrying cable is accidently damaged, or if a fault occurs in electrical or electronic control equipment.

Motors and other electrical machines may also give sparks if they are of a commutator type where a rotating contact repeatedly makes and breaks a circuit. The incendivity of the sparks produced by a circuit depends on the capacitance, inductance, voltage, and current in the circuit, the way in which it is broken, and the shape of the conductors between which the spark occurs.

The most important factor is generally the self-inductance L of the circuit. While the circuit is complete and current I flows, energy is stored in the magnetic field associated with the inductance. When the circuit is broken the magnetic field collapses and the energy $\frac{1}{2}LI^2$ must be released. This energy release helps prolong the spark. The circuit is made safe by dissipating the energy by means of a noninductive resistance or capacitor or by incorporating a closed-circuit secondary winding on the same coil as the inductive component to absorb some of the energy.

The minimum current that will ignite in the absence of these protective measures is a function of the inductor. Widgington (1968) has investigated the design criteria for safe circuits. They show that there is a linear relationship between the logarithm of the inductance and the logarithm of the minimum current for ignition, i.e.,

$$LI^{\alpha} = \text{constant}$$

Calculation from experimental results for gases gave $\alpha = 1.79$ and the constant = 0.00366. If the igniting energy W is calculated from the equation $W = \frac{1}{2}LI^2$, an energy of the order of 0.1 mJ is found for large inductances of approximately 1 H and 0.25 mJ for smaller inductances of the order of 0.3 mH. These are equivalent to currents of 0.04 and 3.9 A, respectively, and comparable to the minimum ignition energy measured with an electrostatic spark.

Sparks may also be produced from the energy stored in a capacitance. The stored energy is given by $\frac{1}{2}CV^2$. Therefore, assuming a 24-V circuit and an ignition energy of 0.2 mJ a maximum capacitance of 0.7 μF is acceptable. An intrinsically safe circuit must therefore not exceed safe limits of voltage current inductions or capacitance.

The dangers of electrical sparks in a gaseous environment is fully recognized and electrical equipment must be either flameproof or intrinsically safe if it is to be used in a designated spark-free area. There are numerous standards and codes of practice, which differ slightly from country to country, that describe precautions to be taken in the presence of a flammable vapor. These are detailed in Appendix A2. Very few of these make any reference to a dust environment. Eventually, British Standard BS 5345, Part 10 will deal with the installation and maintenance requirements for electrical apparatus for use with combustible dusts, but this has not yet been published and at present there is no British standard or European classification specifically dealing with dust atmospheres.

However, information is available in the United Kingdom in publications by the Royal Society for the Prevention of Accidents and in the ICI publication *Electrical Installations in Flammable Atmospheres* which has a section on dust clouds. In the United States and Canada, publications by the Underwriters Laboratory and by the Candian Standards Association (C22.2) No. 25, 1966 do specifically include dust environments as does the National Electrical Code (pp. 502–503).

Two dangers must be defined for electrical equipment to be used in a dust environment. The first is the possibility that a fault will occur, causing a spark which will ignite a flammable dust cloud. The second is the entry of dust into electrical equipment where it causes a fault which leads either to sparking or to overheating. Ingress of a conducting dust into electronic equipment may cause a short circuit, and any dust can prevent adequate cooling. It is less likely that a flammable dust cloud will exist within a piece of equipment than that gradual ingress of dust will cause sparking or overheating which in turn ignites dust deposits in the equipment circuitry itself. A greater stress is therefore put on dust-tight equipment rather than intrinsically safe equipment to be used in a dusty environment. Intrinsically safe equipment is often the best means of protection in a gaseous of mixed atmosphere but, as R. J. Redding (1979) points out, it is not a sufficient criterion in a dusty environment where it is also important that equipment should be dust tight.

Intrinsic safety is achieved by limiting the supply, generation, and storage of electrical energy so that any sparks or hot surfaces produced are unlikely to have sufficient energy to ignite a flammable atmosphere. This limit is applied not only to circuitry but also to cables and terminations. For example, the minimum insulation thickness and mechanical protection of cables is defined and the mechanical strength of terminations is also specified.

A summary of the United Kingdom code of practices relating to intrinsic safety and other related codes are given by G. Tortioshell (1980). Intrinsically safe equipment generally operates on low current and voltage, with a limitation on the size of the capacitance and inductance of the circuitry in which energy is stored.

The rapid development of microelectronics has allowed more control and test equipment to be made intrinsically safe with ease, and high-current-mains operated equipment is being replaced. Electronic barriers can be fitted between intrinsically safe equipment in a hazardous area and associated intrinsically unsafe apparatus that is kept in a safe area.

Alternatives to intrinsic safety, where it is impossible to limit the energy effectively, are to pressurize the equipment so the atmosphere cannot enter or to fill the equipment with an inert powder or encapsulate it in epoxy. These methods are described in British Standards publication

B.S. 5345 and B.S. 5501. (B.S. 5001 is equivalent to the Cenelec European Standard EN 50 020).

A dust-tight enclosure is defined in American Standards NE 7A, standard IC-1, Industrial Control. It is a substantial mechanical contruction with gaskets or the equivalent to exclude dust and has no holes. The conduit entrance is by fine-tapped threads with a minimum of three threads engaged or by a gasketed hub. The fastening of doors and covers is also specified in a way that ensures they are dust tight. It is important that the temperature on the surface of a dust-tight enclosure cannot rise and American specifications require a temperature check.

High-voltage equipment such as electrostatic powder-spray guns are not intrinsically safe but may be protected by ensuring that there is a low-current limit. This may be achieved by a high series resistance in the output (Figure 2.2). For example, if the spray gun has an operating voltage of 80 kV and a resistor of $10^9\ \Omega$ is fitted immediately before the output, then a short to the output still leaves $10^9\ \Omega$ in the circuit to ground. The maximum current I that can flow is then given by Ohms' law $I = V/R$

$$I = \frac{8 \times 10^4\ \text{V}}{10^9\ \Omega} = 8 \times 10^{-5}\ \text{A}$$

This current is insufficient for ignition. The effective voltage after the resistor is very low because all the voltage is dropped across the resistor. The capacitance of anything that follows the limiting resistor must also be low so that the maximum charge stored is insufficient for an electrostatic spark.

Current trips may be fitted as an alternative to a limiting resistor so that when the current exceeds a fixed level it switches off. Current trips must be fast operating and fail-safe.

A detailed standard for motors and generators to be used in a dust environment, but which are not intriniscally safe, has been published by the Underwriters Laboratory in the United States (1973). The aim of the standard is to limit the temperatures to which surfaces may heat and prevent the ingress of dust into the moving parts of machinery or the

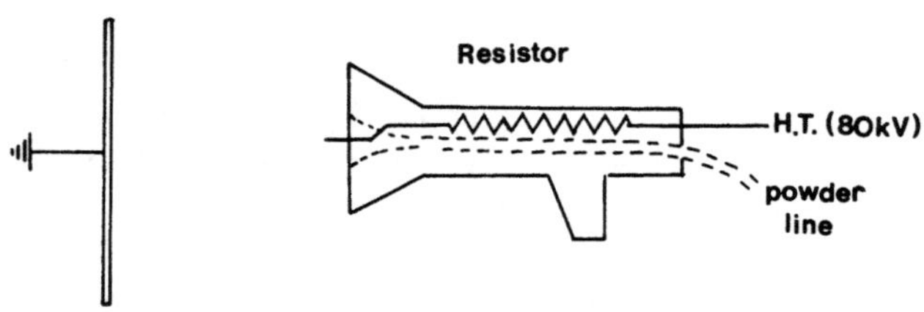

FIGURE 2.2. Electrostatic powder-spray gun.

electrical circuitry. The motor or generator must be totally enclosed and the clearance on the shaft and other necessary inputs to the enclosure are closely specified. Temperature-limiting devices must be fitted precisely to limit the temperature of external surfaces.

British and European standards have no special regulation for dusts, but motors and generators have three types of protection:

(i) "N." Protection such that in normal operation the motor is not capable of igniting a surrounding explosive atmosphere and a fault capable of causing ignition is not likely to occur.

(ii) "E." Additional measures applied to electrical equipment so as to give increased security against the possibility of excessive temperatures and of the occurrence of arcs and sparks. It applies only to electrical equipment, no part of which can produce sparks or dangerous temperatures in normal service. This type of motor can be used in Zone O and I (U.K.) or Division I areas (U.S.).

(iii) Flameproof enclosures. An enclosure for electrical apparatus that will withstand an internal explosion of flammable gas or vapor which may enter it without suffering damage or communicating the internal combustion to the external flammable gas or vapor through any joints or structural openings in the enclosure. This is very similar to the American National Electrical code. Class I for explosion-proof motors.

The National Electrical Code also has a Class II specification dust ignition-proof motors for which there is at present no European equivalent. A dust ignition-proof enclosure is defined as one which will exclude ignitable amounts of dusts or amounts that might affect performance or rating and which will not permit arcs, sparks, or heat otherwise generated or liberated inside the enclosure to cause ignition of exterior accumulations or atmospheric suspensions of a specified dust on or in the vicinity of the enclosure.

Case Histories—Electrical Sparks

U.K. 1978

A spark from an electrical junction box ignited dust that was being swept away. The damage amounted to £100,000. This incident clearly shows both the importance of good housekeeping and cleanliness and the necessity of ensuring that suitably protected electrical equipment is used in all areas where there can be a flammable dust.

U.K. 1977

Water seeping through the roof of a single-story building caused an electrical busbar to short circuit and ignite combustible materials nearby. The sprinkler system operated for only a short time before the supply of electricity to the pump failed which was the only water source supply to the sprinklers. The damage cost £340,000.

Plastic Injection-Molding Plant: U.S. 1978

The building, used for plastic injection-molding operations, was of noncombustible construction. Part of it was also used for storage of plastic (polystyrene) molding pellets and for offices. The pellets were stored on skids in 1000-lb cardboard containers. In the same section of the building, a 16-in. × 20-in., 20 hp plastic grinder was housed in an 8 × 8 × 8-ft^3 wood-frame enclosure insulated with fiberglass. Plastic-dust accumulations inside the enclosure were reportedly cleaned out daily.

There were wet-type automatic sprinklers throughout the building, except in the grinder enclosure.

At 8:00 am, after the weekend shutdown, the grinder operator pressed the "start" button inside the enclosure. There was an electric arc from the button. Plastic-dust accumulations inside the grinder enclosure ignited, and a flash fire ensued. The foreman, public fire department, and plant emergency organization were notified at once.

Five ceiling sprinkler heads in the immediate area outside the enclosure activated 5 min later. The sprinklers, together with the plant fire department, extinguished the fire before the public fire department arrived at 8:15 am.

Fire damage was confined to the grinder and enclosure. There was some smoke and water damage to the building, equipment, and storage in the immediate vicinity. Loss was estimated at $25,000.

Dust Explosion Started by Flammable Vapor Explosion: U.K. 1971

A severe explosion and fire occurred in a Glasgow dockside processing plant for extracting oil from soya beans. A boilerman noticed a strange smell in the underground pump house which was identified as hexane. When he attempted to switch off the electrically driven pump, arcing from the nonflameproof switch caused a minor explosion of hexane vapors. This initiated a dust explosion in the tunnels adjoining the pumphouse. These tunnels were used to feed the silos with grain and to take grain to the process area by means of conveyor belts, and there was at least 6 mm of dust in them. The dust explosion ran along the passageways from the pumproom, along the quayside, blowing out a thin wall between two of the tunnels and hurling bricks 183 m down one tunnel and along the quayside. The explosion blew out each manhole cover as it reached it, and wrecked the conveyor belt machinery as it passed.

At the same time the explosion passed beneath the four silos into the sump beneath them, breaking through into the space between them. Dust was stored here and another explosion occurred. Fed by further dust, the explosion passed under a fifth silo and vented itself through a 2 × 2 m^2 manhole.

Later investigations revealed that a tank containing hexane was leaking; the hexane had percolated into the silo and eventually penetrated as far as the pumphouse.

Three employees and five visitors to a ship at the quayside were injured, though the latter were not detained in a hospital. Damage included the four storage silos, metal trunking, conveyor belts and associated machinery, part of a concrete floor and of the quayside, and 20 manhole covers. The berthed vessel was slightly damaged by falling debris. Estimated loss was £260,000.

2.4.2. Frictional Impact Sparks

When metal strikes against stone or rock a spark may be produced. This spark can, under some circumstances, ignite a flammable atmosphere. Most of the research into this phenomena has been performed for the coal-mining industry where cutting tools create showers of sparks and methane gas may be present. Very little work has been carried out on the ignition of dust atmospheres by this mechanism, and as yet no one has succeeded in igniting a dust cloud in the laboratory by a impact spark unless a chemical reaction such as the thermite reaction is present or the dust is particularly sensitive to ignition. Nevertheless, in recent years a number of industrial dust explosions have been attributed to the impact of metal on metal or metal against stone, and the question of ignition dusts by this mechanism must be considered.

Since a flash of light is observed during impact, there is a tendancy to associate impact sparks with electrostatic sparks. This is misleading since the source of ignition in an impact spark is not the heat generated by the rapid ionization of the air as in an electrical spark, but the heated surface created by impact. The "spark" is in fact a very hot fragment of material. Ignition is thus more closely related to ignition by a hot surface than by an electrical spark.

It is not possible to consider an impact spark in terms of energy. When a 0.5-kg object falls a distance of 2 m the total energy dissipated in the impact is 10 J. This is considerably greater than the minimum ignition energy of many dust clouds, but most of the impact energy is dissipated as heat in the metal or stone striking surfaces and only a tiny fraction is transferred to the dust cloud. This amount cannot be easily measured or estimated.

The ignitability of an impact spark must be measured by finding the gas mixture at initial ignition and then finding the minimum ignition energy of this mixture. Unfortunately there is no reason why the method by which heat is transmitted from a small hot particle or hot spot on a surface to a gas should be the same as the way it is transferred to solid dust particles. The measurement in gases therefore provides a general comparison rather than a direct measure of ignitability of impact sparks applicable to dusts.

Impact causes heating in two distinct ways. There is a direct transfer of the kinetic energy of the impact to heat at the impact point. The temperature rise over the impact area will depend on the thermal conductivity of the two bodies and their specific heat. There may also be heat generated due to exothermic oxidation of the surface, initiated by the temperature rise due to the impact. This is highest for the easily oxidized materials such as magnesium, aluminum, and iron.

The temperature rise at the impact point is usually governed by the more-easily-oxidized material and this and other evidence suggests that the principle heating, at least for oxidizable materials, is due to oxidation rather than direct conversion of the impact energy (Bowden, 1954).

Calculation shows that oxidation of 1 g of iron will produce 6.6 kJ of energy, and oxidation of copper 1.3 kJ. Since copper also has a much greater thermal conductivity, it can be seen that the temperature rise due to impact on iron will be much higher than for copper (Navord Report 5205).

A spark created by friction is a very hot fragment of material knocked off by impact. Again the very high temperature of the fragment is usually due to oxidation of the surface. The most extensive work on ignition by friction has been carried out by Bowden and his group at Cambridge (Bowden, *et al.*, 1947; Bowden, 1949, Bowden and Thomas, 1954; Bowden, 1954). They were interested primarily in the ignition of explosives such as nitroglycerine; nevertheless, many of their general conclusions concerning the effects of the characteristics of hot spots created by rubbing and impact apply equally to dust explosions.

Figure 2.3 shows the measured maximum temperature produced by steel sliding on glass as a function of speed and load measured in Bowden's work (1954). (It is possible that the curves are slightly affected by softening of the glass above 800° C). The duration of a hot spot was found to be a few milliseconds in which time the slider had moved ~ 4 mm and was no longer in contact with the hot spot. These measurements are therefore applicable equally to rubbing contacts and a single glancing blow. It can be seen that temperatures up to the melting point of metals could easily be produced by sliding at moderate conditions of load and speed with materials such as rock, steel, and glass.

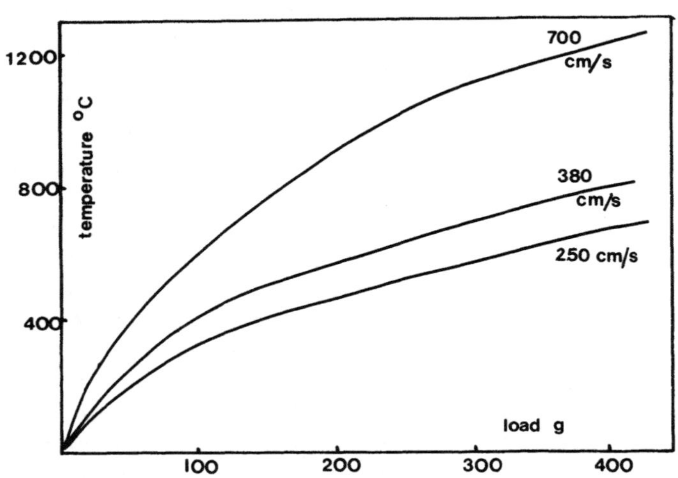

FIGURE 2.3. Variation of maximum temperature of hot spots with load and sliding speed for steel sliding on glass.

A metal that oxidizes easily in air, such as aluminum, could produce extremely hot spots that had a temperature above the melting point of the metal. The heat released is then due to exothermic oxidation of the surfaces initiated by the original friction heating.

The characteristics of the hot spots produced by impact have been studied more recently by Blickensderfer (1975). Using an optical pyrometer he measured the temperature produced by the impact of a cutting tool attached to a fly wheel and then calculated the expected temperature profile. Figure 2.4 shows an oscilloscope trace of the temperature of the hot streak as a function of time. The hot area was the same width as the striking edge of the tool while the length depended on the striking velocity. Elevated temperatures were found to exist over areas up to approximately 1 cm² for several milliseconds.

In summary, the experimental work shows that sliding at speeds between 1 and 7 m/s and loads between 100 g and 10 kg produce temperature rises ranging from 300°C to 1400°C over contact areas of a few square millimeters in hot spots representing actual contact points of $\sim 10\mu$ in diameter. The temperature will rise still further, above the melting point of the impacting materials, if oxidation can occur. The duration of the hot spot was measured by Bowden and Blickensdefer to be a few milliseconds. In that time the slider had moved away from the heated area, so the measurements also apply to a glancing blow.

Theoretical treatments of the temperature rise during friction have been developed by Block (1937), Jaeger (1942), Holm (1948, 1952), and Archard (1952). The basic philosophy of the calculations is to assume the

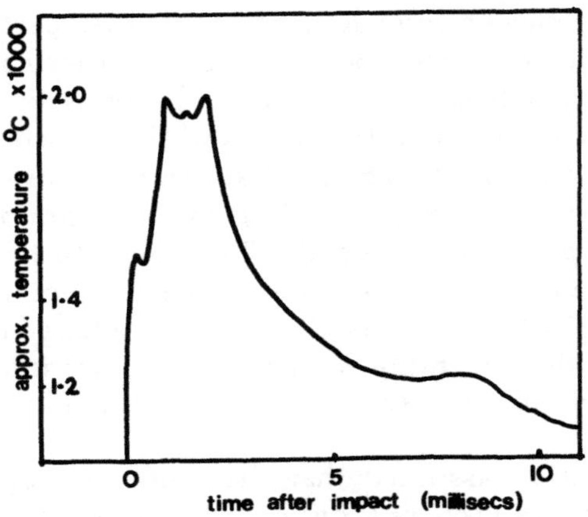

FIGURE 2.4. Hot-streak intensity for impact of steel on sandstone. The area examined was 1 cm in diameter.

energy of friction is converted to a heat source at the area of contact and to calculate the rate at which heat is dissipated within the bodies. The temperature rise thus depends on the load, velocity, and coefficient of friction as well as on the conductivity and specific heat of the materials.

There is no doubt that the temperatures produced during either impact or sliding will exceed the ignition temperatures of many dusts. However, it is necessary also to estimate whether it is possible for the heat produced to be transferred to dust particles either in a cloud or in a layer on the impacted surface. The work on gases shows that a minimum hot area of a few square millimeters is required for a gas to ignite (Rae, 1956; Powell, 1969) but the equivalent parameter for powders is not known.

The mechanisms of heat transfer from a hot surface to dust particles in a cloud have not been investigated. The problem is similar to that of the transfer of heat from individual particles which sets up the chain reaction causing an explosion, except that the temperature of the hot spot on the impacted surface falls by conduction within a few milliseconds.

When there is a dust layer between two impacting surfaces it can be assumed that some of the mechanical energy of the impact is converted directly to a rise in temperature of the dust, as well as heat being transferred indirectly to the dust from the heated solid surfaces. However, this problem has been treated only by those interested in powdered explosives which are frequently intentionally ignited by impact. Although the ignition data are not relevant to normal dusts, the fundamental data on temperature rise are useful. Chaudhri (1976) has calculated the temperature rise when an indenter is impacted onto a compacted explosive powder sample. From experimental studies of the deformation of powder compacts, he found that particles of powder became firmly attached to the indenter. Since the thermal conductivity of the particles is much less than that of the impacter, the maximum temperature rise will occur at the surface of the particles attached to the indenter. The calculations made by Chaudhri therefore assume that the particles attached to the indenter rub against a continuous plane surface provided by the particles in the bulk. The temperature rise as a function of penetration of the indenter is shown in Figure 2.5 and reached ~100°C. The validity of the calculation was evaluated by choosing two explosives, one with an ignition temperature below the level predicted by Chaudhri's theory and one above. The ignitions obtained supported the predictions.

This calculation assumed that the indenter stopped within the compacted material and did not consider a thin layer of powder on a metal surface penetrated by the indenter. But it can be seen from the figure that the dust on the impacter surface reached the maximum temperature

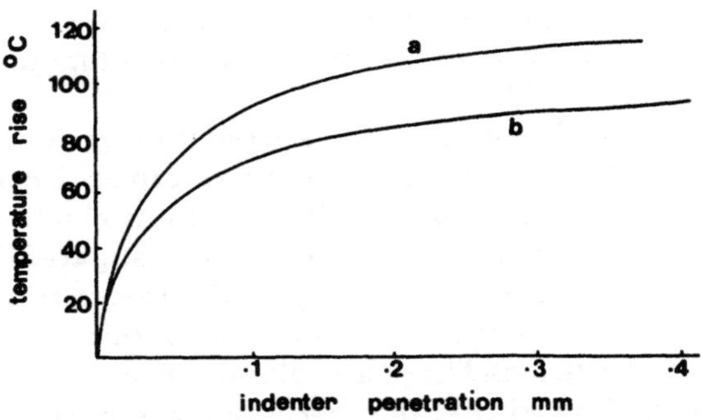

FIGURE 2.5. Calculated temperature rise as a function of indenter penetration for two conical indenters of different mass with the same initial kinetic energy; (a) mass 0.145 kg, velocity 0.5 m s; (b) mass 0.29 kg, velocity 0.353 m s.

with penetration of a fraction of a millimeter and eventual impact with the surface can serve only to increase the temperature rise.

Other studies of the initiation of explosives in powdered explosives have been carried out by Heavens and Field (1974), Field and Winter (1975), and Heighway and Taylor (1958).

The mechanisms of heating in a dust layer during impact is still a matter of debate. Apart from direct frictional heating discussed by Chaudhri, heat may also be produced by adiabatic compression of the gas trapped between the particles, viscous heating of material that melts and extrudes between the impacting surfaces, and localized adiabatic deformation of thin layers upon mechanical failure.

Winter and Field believe the major heating effect occurs during plastic deformation of the particles. The mechanism by which heat can be produced during impact onto dust covered surfaces are extremely complex. However, as a first step the initiation of dust explosions can be treated by assuming that the mechanical energy of impact of dust explosions is converted into heat spread over a reasonable contact area. This exercise has not been carried out for dusts that are not designed as explosives, and heat transfer from individual hot particles knocked off a surface to a dispersed dust cloud has also not been treated in the general literature.

2.4.3.1. *The Thermite Reaction*

The thermite reaction is the exothermic reaction that occurs when a metal with a high affinity for oxygen comes into contact with the oxide of a metal with a lower affinity for oxygen. The most common combination is aluminum and rust. The reaction, which may be initiated by impact or by heat, reduces the iron oxide and produces aluminum oxide and considerable heat.

Only a smear of aluminum need be present to ignite either gases or powders (Grice, 1932). The ignition mechanism can be either the burning aluminum at the impact point or particles of burning aluminum projected from the impact. Aluminum paints have been investigated (Kingman *et al* 1952) and some have been found to give the thermite reaction. The danger depends on the paint formulation. This is constantly changing so the safest solution is to avoid the use of aluminum paints. Gibson, Lloyd, and Perry (1968) showed that the thermite reaction could be initiated when an aluminum smear on rusty iron was hit with an implement made of any material including a "nonsparking" hand tool.

Magnesium and aluminum were shown by Bowden and his co-workers to give higher-temperature hot spots because of their affinity for oxygen, even when the thermite reaction was not involved, and this was confirmed by Blickensderfer (1975) who showed that easily oxidized metals ignited gases more readily during impact.

2.4.3.2 Ignition of Gases by Friction and Impact—Laboratory Tests

A large body of work has been carried out concerning the ignition of gases by impact and rubbing. The majority has been specifically aimed at the coal-mining industry where cutting tools must be used and methane vapor may be present. The ignition source may be either the hot surface produced by the impact, or the projection of hot and possibly reacting particles into the atmosphere (the so-called impact spark).

Although the results of work on gases are not directly applicable to dust clouds, useful information is obtained on the conditions which produce sparks that ignite gases most easily. Rae (1956) found the minimum size of a heated metal that would ignite a gas when set in the side of an explosion box. At 1250°C a 6-mm square was required but at 1600°C the size was reduced to a 2.5-mm square. Coal gas and pentane ignited more easily than methane.

No similar work has been carried out on powders. Experiments firing small high-speed balls (1–2 g) at rock targets in a methane atmosphere gave no ignitions of methane, but hydrogen ignited readily. Glancing impact was shown to be much more effective than normal impact, which is to be expected since a minimum size of heated area is required. It has been shown that the speed of impact affects the ignition probability more than the impact energy because the higher speed increases the size of the hot area rather than because the temperature is higher or last any longer. The oxidation behavior of the metal involved in the impact is also important with more easily oxidized materials igniting gases more frequently.

A considerable number of tests on gases have been carried out with simulated cutting tools or with simulations of hand tools and equipment falling. This has been reviewed by Powell. The results can be summarized as follows: (i) Gases and vapors (similar to propane or petroleum vapor) ignite with difficulity from impact between steel and steel or steel and rock. (ii) Hydrogen-type gases ignite easily in these circumstances. (iii) Impacts involving aluminum or magnesium will cause ignitions of propane and similar vapors easily.

The only extensive published study of dust ignitions is the work of A.R. Morse. He failed to obtain ignition of a grain dust cloud in a small chamber using a stone striking against a file, but easily ignited the dust by impact of a cerium iron flint. A larger chamber (1 ft high, 1 ft in diameter) was used for tests with direct simulations of falling objects from heights of 25, 40, and 85 ft. No ignitions were obtained but very bright sparks were observed which in some of tests might have been small flame kernals that did not propogate. Brighter sparks were obtained more often when a steel on steel impact was used than when the falling object was rock. Every attempt was made in these tests to ensure the ignition of a dust cloud was being studied rather than a settled layer, and the chamber was vibrated every 15 s to try to keep the dust in circulation. Ignition was obtained on only the second attempt when a rusty steel anvil was struck by an object weighing 5 lb and made of magnesium containing 10% aluminum. Gibson, Lloyd, and Perry used a spring-loaded hardened-steel hammer (1.4 kg) which was accelerated rapidly and struck a nearly tangential glancing blow onto a steel block. Caps of different metals were fixed to the hammer and the steel block was rusted for tests of the thermite reaction. The thermite reaction produced a bright white flash whereas nonthermite sparks appeared red. Gibson and co-workers tested both gases and organic powders (such as used in the dye industry).

For the dust tests a small pile of dust was placed on the steel block and was dispersed by the hammer (Figure 2.6). It was found that dust ignitions could occur when the thermite reaction took place but no ignitions were obtained from nonthermite sparks. No attempt was made,

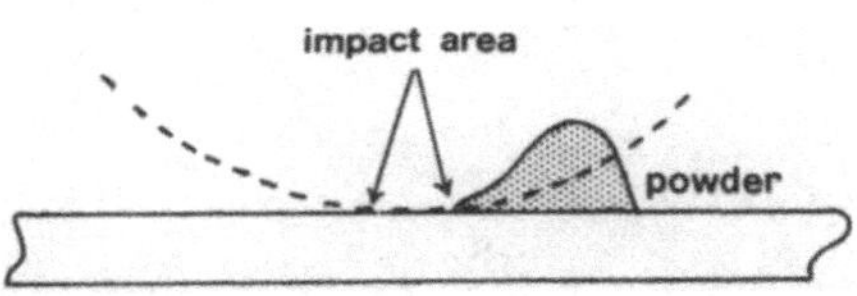

FIGURE 2.6. Schematic diagram test for ignition of a powder by an impact spark.

however, to control the dust-cloud density in these tests so there would be no validity in the statement that nonthermite sparks cannot ignite dust clouds. Once again there was no dust layer between the impacting surfaces.

The United States Bureau of Mines has investigated ignition of mixtures of coal dust, air, and methane, simulating impacts between steel and minerals which might be present during pneumatic transport of coal (Kelley and Forkner, 1976). Coal dust in air alone was not ignited by abrasive impacts but addition of as little as 1-vol % methane gave ignition. No ignitions occurred with impacts of steel on limestone even when methane was present and spark showers were produced. There were no ignitions at impact velocities below 900 ft/min. The ignitions site was found to be the friction-induced hot spot on the rod rather than the sparks produced by the impact.

Laboratory studies of dust ignitions by impact suffer from two difficulities. It is necessary to provide a controlled dust mixture and keep it suspended at an explosive concentration for the period of the tests, and it is very difficult to produce reproducible impact sparks of known energy. However, laboratory tests so far have not succeeded in igniting a nonsensitive dust by impact unless the thermite reaction was involved. Sensitive dusts such as aluminum, zinc, magnesium, and sulfur have, however, been ignited without difficulty (Brown, 1941).

2.4.4. Nonsparking Hand Tools

The temperature of the hot spots produced during friction or of the hot streak produced by impact depends on the energy dissipated in the contact, the thermal conductivities of the impacting bodies, and the surfaces' tendency to oxidize. In many cases the exothermic oxidation reaction is the chief source of heat during and immediately following impact. Nonsparking hand tools are generally made of materials that do not tend to oxidize or whose oxidation reaction produces little heat and that also have a high thermal conductivity so that the heat created is rapidly dissipated. Nevertheless, nonsparking tools can still produce impact sparks that can ignite flammable vapors.

The use of nonsparking hand tools is recommended for use near highly flammable gases such as hydrogen but not generally for propane-type gases or for dusts.

2.4.5 The Industrial Position

The laboratory evidence indicates that dust ignition by single impacts not involving the thermite reaction are improbale. The statistics of accidents are difficult to interpret because full detailed investigations of the ignition source are usually only carried out where damage is severe enough to halt production, and in these cases it is difficult to find out whether a piece of broken metal caused the explosion or was caused by it.

The number of ignitions attributed to single-impact is in any case low and in view of the unreliability of accident reporting not statistically significant.

Although there is insufficient data to be sure whether dust clouds will ignite from the single impact of a falling object, it is probable, both from the statistics and from theoretical considerations, that processes involving continuous rubbing—particularly where dust may be trapped between the rubbing surfaces—can cause overheating and lead to a dust ignition. Tramp metal that finds its way into grinding or mixing machines and the continuous impact of scraping metal surfaces are also well established causes of ignitions in industry.

Case Histories—Impact Sparks

Foundry Manufacturing Aluminum and Magnesium Aircraft Parts: U.S. 1976

Although the exterior walls of the foundry were primarily of concrete blocks, there were many wood-on-wood frame partitions within the building, separating the foundry's internal operations (i.e., casting, heat treating, and grinding). The building had an ordinary combustible, board-on-joist roof.

At approximately 9:15 am, an employee was grinding one of the steel dies to correct an alignment problem with the parts cast in the die. The sparks given off from this operation ignited magnesium-dust accumulations on the floor and walls of this area. The fire quickly spread across metal-dust accumulations on some wooden shelves and along the underside of the board-on-joist roof. Employees attempted to use a wheeled, 300-lb dry chemical extinguisher, but the extinguisher was inoperable. By the time they were able to obtain a second similar extinguisher, the fire was out of control.

The Fire Department was summoned and responded at 9:20 am. Within 3 minutes after fire fighters arrived, the entire roof of the building was enveloped in flames. Three alarms were sounded and eight engines, two ladder trucks, one aerial platform, and 100 men responded. The fire was brought under control by 11:30 am and extinguished 2½ h later. Fire officials estimated that they had used 1,100,000 gallons of water during their operations.

Fire investigating officials theorized that the rapid spread of this fire was caused by lack of adequate fire divisions, absence of automatic sprinklers, and inadequate housekeeping to prevent excessive buildup of combustible metal dust.

This plant had experienced a similarly destructive fire 12 years earlier.

The building had an automatic fire-detection system, equipped with fixed-temperature heat detectors. The system was wired to provide a proprietary alarm, but it did not operate during the incident.

Probably an important factor in the $1 million property damage loss was the failure of the first large, wheeled fire extinguisher to operate properly. The foundry's extinguishers

were checked on a monthly basis and serviced yearly. Apparently the monthly inspection entailed discharging a small portion of the dry chemical. The faulty extinguisher had been checked only 3 days before the incident. However, the combination of a blockage in its hose and a reduced charge were determined to have caused its malfunction.

2.5. SECONDARY IGNITION

Many major explosions of dust clouds are technically known as *secondary explosions*. That is, the ignition source of the major explosion is heated powder or burning fragments from an initial small fire or explosion with the plant. Often the dust cloud that explodes in the secondary explosion is not present until dust is raised by the primary ignition. Damage is often vastly decreased if small fires and explosions, which are extremely difficult to eliminate totally from all areas, are prevented from spreading and initiating further explosions.

Two types of precaution can be taken. First, very strict housekeeping can be instituted so that there are no dust layers to propogate a smouldering fire or to be raised by a small explosion. Second, checks can be applied both automatically and manually to ensure that fires cannot gain a hold even in areas where there is no dust cloud.

It is advantageous if areas can be isolated from each other so that dust which may tend to build up in one section is not spread throughout the plant. This principal can also be applied to the system itself and interconnecting sections should be minimized as much as possible. For example, a dust-extraction system that serves the whole plant, may, if poorly designed, be able to spread an explosion throughout the plant. Separate extraction systems for different parts of the system can limit damage.

In an ideal system where areas containing a dust cloud are protected by venting suppression or inerting and all other parts of the plant are scrupulously clean, damage due to an explosion should be minimal.

3

Static Electricity

3.1. INTRODUCTION

The phenomenon of charge separation accomplished by rubbing was discovered by the ancient Greeks; the Greek word for amber, the material in which charging by rubbing was first detected, is the origin of the word *electron*. In spite of these early beginnings the details of how and why charge transfer occurs are still not fully understood today. Electrostatic phenomena are strongly affected by minor changes in the surfaces involved, making results unpredictable; this situation has led to a tendency to regard electrostatics as rather more an art than a science. This view is reinforced by difficulties of measurement. Many of the instruments used to measure static charge and voltage affect the quantity they are measuring, so an allowance must be made for instrumental effects.

In spite of these apparent difficulties, it is possible to quantify electrostatic effects reasonably well, and it is the intention of this chapter to remove some of the mystery surrounding the subject.

There are five quantities fundamental to an understanding of electrostatics. The most basic is the *electric charge* that is transferred to a material, usually by friction. When an object is charged it exerts a force on any other charged object. It is then said to have an *electric potential* or *voltage* V. The rate of change of voltage with distance or potential gradient is the *electric field* E. The potential reached by an object having a charge q depends on its *electrical capacity* C. The higher the capacity the more charge is needed to achieve a given potential.

Finally the rate at which charge leaks away depends primarily on the *electrical resistance* R between the stored charge and ground. The relationships between these quantities and other fundamental relationships are listed in Appendix A1.

3.2. DEFINITIONS AND RELATIONSHIPS

3.2.1. Electric Charge

3.2.1.1. Frictional Charging

The S.I. unit of electric charge is the coulomb (C) where the electron has a charge of 1.6×10^{-19} C and where a current of 1 A = 1 C/s. During rubbing or contact, charge is transferred from one material to another. The maximum charge that an isolated insulator surface can hold is 3×10^{-5} C/m^2 (equivalent to approximately 2×10^{14} electrons). Above this level minute surface discharges immediately occur to bring the charge back down to the maximum. Although this charge level creates a high potential the charge itself is not great. There are of the order of 10^{20} atoms in every square meter of a solid's surface; thus, only two atoms in every million have gained or lost an electron when an insulating surface is charged to its maximum level. It is clear that surface impurities can play a large part in electrostatic phenomena, and very small changes in the surface content, perhaps by only a few parts per million, can greatly affect electrostatic measurements.

It is partly for this reason that it is still impossible to accurately predict how any given material will charge. It is known that a material will achieve a greater charge in direct relation to the amount of work done upon it. Thus micronizing and pneumatic conveying of powders produce more charge than sieving or pouring. However, a comparison between different materials is difficult to make.

To some extent it is possible to predict the polarity of charge from an experimentally determined triboelectric series. Materials are arranged in a table such that each material charges samples below them in the table to a more negative potential. Thus, for example PTFE is negatively charged by most other materials and is at the bottom of the table, while nylon, which usually becomes positive, is near the top.

Agreement between triboelectric series published by different laboratories is not good and some research has shown cyclic situations that make it impossible to find an unambiguous ordering. Real surfaces of relevance to industrial situations cannot be defined sufficiently well to enable published triboelectric series to be used as anything other than a general guide to the probable polarity which will be produced on a material by friction.

Laboratory tests may be devised to evaluate powder charging but none of the tests tried so far have given a clear indication of how much charge can be acquired in an industrial situation. The best that can be achieved is a comparison between similar products in order to identify which is the most likely to produce high charge levels in practice. The

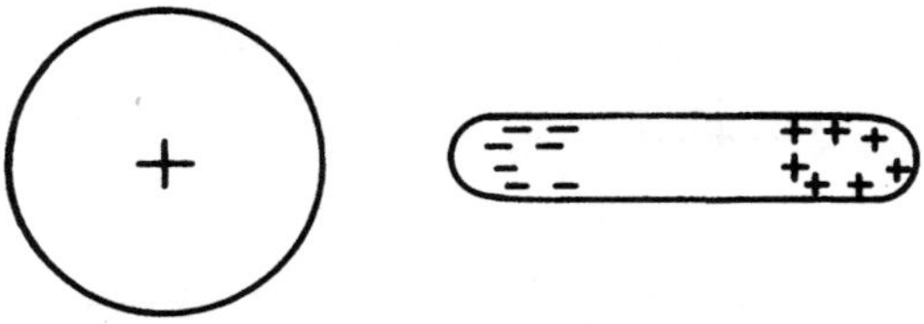

FIGURE 3.1. Polarization of a conducting object.

ways in which charge is produced on powders by friction is discussed further in Chapter 4.

3.2.1.2. *Induction Charging*

Charge can be transferred from one object to another without direct contact between them. This is known as *induction charging* and is fundamental to an understanding of many electrostatic measurement techniques.

When a conducting article is placed in an electric field the internal charges are free to move. Since unlike charges attract, the article polarizes so that one side is effectively positive and the other negative. Similarly, if a charged object is brought close to an isolated conductor the opposite polarity charge is attracted as illustrated in Figure 3.1. When the conductor is touched with a grounded rod only half the charge flows to ground—one polarity remaining effectively bound by its attraction to the

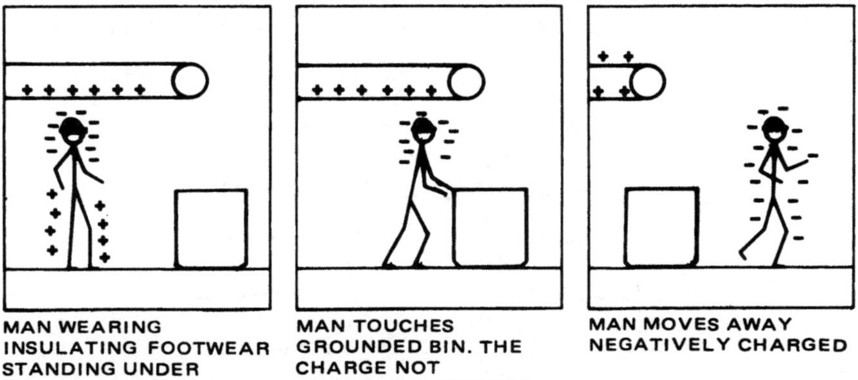

FIGURE 3.2. Example of a charged object brought in close to an isolated conductor, and the resulting attraction of an opposite polarity charge.

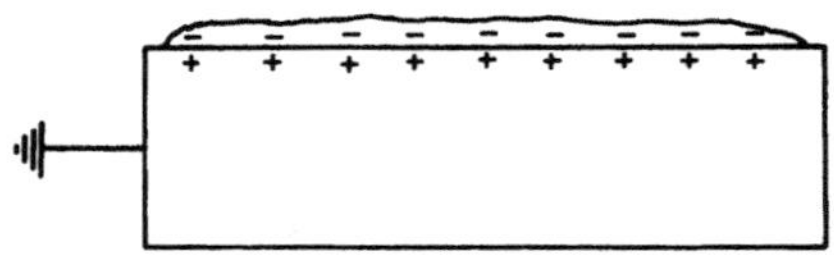

FIGURE 3.3. Charged powder and its image in grounded metal surface.

other object. When the two are separated the conductor is left with a net charge. It can be seen that this mechanism can lead to the accumulation of a high charge on personnel working in the vicinity of an electrostatic charge—for example, near an insulating plastic belt or a highly charged insulating powder—even though there is no direct contact with the charged object (Figure 3.2).

The charge induced on the surface of a conductor due to the presence of an external charge is often known as the image charge. It is opposite in polarity to the charge which creates it and may be treated mathematically as its mirror image within the surface. For example, a charged layer of insulating powder on a metal surface is balanced by its image charge so the effect of the charge above the surface is extremely small. Therefore, little repulsion is experienced from a charged powder layer on a metal surface. (Figure 3.3).

3.2.1.3. Corona Charging

An isolated metal or insulator surface will charge if free ions of one polarity predominate in the air and there is an electric field. Figure 3.4

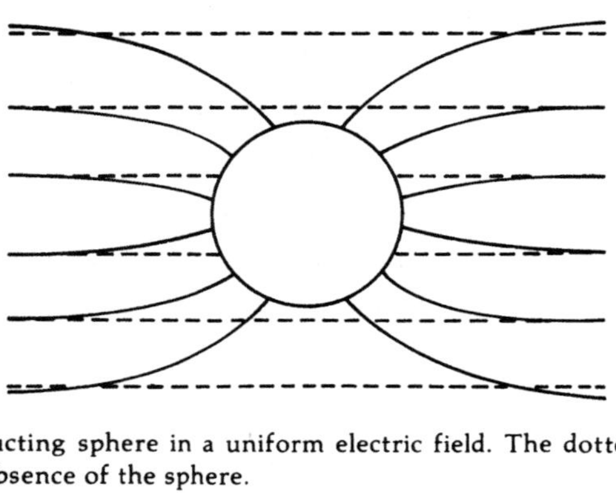

FIGURE 3.4. Conducting sphere in a uniform electric field. The dotted lines show the applied field in the absence of the sphere.

shows a conducting sphere in a uniform electric field. The difference in dielectric constant between the air and the particle causes a distortion of electric field so that charge is attracted to the sphere as described in Section 3.2.3. A limiting charge is reached when the repulsive force due to the charge collected balances this effect. The maximum charge that can be achieved has been calculated by Pauthenier (1932) and is given by White (1963) as

$$q_{\max} = 4\pi\epsilon_0 a^2 Ep \tag{3.1}$$

where a is the radius of the particle, E the field in which it is charged, and p a constant with a value of ~2–3 depending on the dielectric constant of the sphere.

Ions of one polarity seldom predominate in nature but in industry a corona discharge is frequently used to apply a known charge so that the motion of particles can be controlled by electrostatic forces; for example, in electrostatic powder coating, electrophotography, and electrostatic precipitation and separation.

3.2.1.4. Liquid Charging

All liquids, except very highly insulating materials, contain polar molecular groups. At any boundary between liquid and solid or liquid and gas one polarity of ion predominates. Ions of the opposite polarity are loosely attracted to this boundary layer but are more diffusely distributed. A double layer is thus formed (Figure 3.5).

When the liquid flows past the surface the tightly bound charged layer remains while the diffusely distributed charge is carried by the flowing liquid. Similarly, if a surface is broken up or atomized the droplets formed will carry a charge from the surface layers. The charge formed by these processes increases in direct proportion to the increase

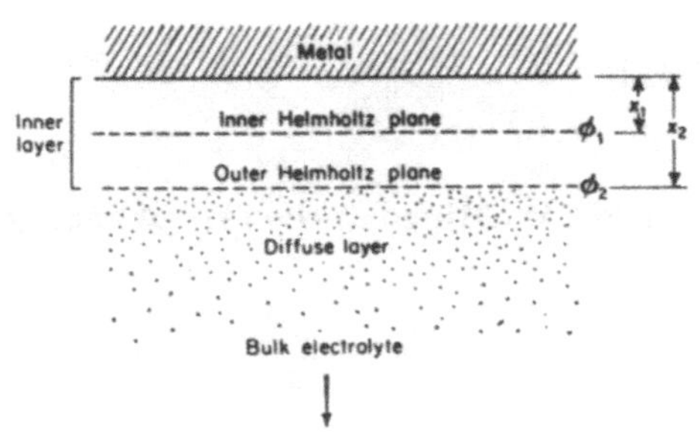

FIGURE 3.5. Double layer formed at the interface of a liquid and a metal pipe.

in the liquid's velocity. It also depends on the number of polar groups present, i.e., on the conductivity of the liquid. Highly insulating liquids do not charge well because of the low number of polar groups present whereas highly conducting liquids leak charge too rapidly to store it. The danger region therefore lies in the intermediate region of conductivity, between 1×10^{10} and $2 \times 10^{12}\ \Omega \cdot \text{m}$.

3.2.2. Electrical Potential

The S.I. unit of potential or voltage is the volt. In electronics or power engineering an electrical potential is usually supplied by a battery or by a mains supply. It is therefore fixed, and charges flow in conductors under the influence of this potential. In electrostatics a charge is created and usually cannot flow. The charge is therefore fixed and the measured potential depends on the electrical capacitance where the charge is stored. The voltage V, the charge q, and the capacitance C are related by the equation

$$V = \frac{q}{C} \tag{3.2}$$

This may be taken as the definition of capacitance. Frequently, it is not convenient to measure charge directly and the voltage is measured instead. The charge is then found from this equation. If a charged surface is moved from one position to another its capacitance may change; therefore, if the charge is fixed the voltage must change. A measurement of voltage without an indication of capacitance is meaningless. This variation of the potential is the source of much of the confusion arising with electrostatic measurements. It can be seen that it is a direct consequence of the fact that the charge is static. If charge is treated as the fundamental quantity on which the voltage and electric field depend much of the confusion is eliminated.

3.2.3. Electric Field

3.2.3.1. *Basic Definitions*

The electric field E is defined as the rate of change of potential with distance, i.e.,

$$E = -\frac{dV}{dx}. \tag{3.3}$$

For a parallel-plate geometry the field is uniform and this becomes simply

$$E = -\frac{V}{d}. \qquad (3.4)$$

where V is the potential difference between the plates and d their spacing.

The electric field is often visualized in terms of imaginary lines representing the path along which a free charge would move. Field lines are always perpendicular to the lines joining points of equal potential and can be considered to begin and end on a free charge. The field around a charged object is directly related to the amount of charge on it. Measurement of the electric field can be used to find the charge when direct measurement is impossible.

The use of equipotentials to predict charge movement can be illustrated using the example of a conducting sphere in a uniform field (Figure 3.6). A conductor must necessarily be at a uniform potential. (If it were not, charge could flow within it until the potential was balanced.) The two electrodes between which the uniform field is applied must also be equipotentials. Every position must have a unique potential—potential lines cannot cross and therefore must follow the direction indicated in Figure 3.6.

Field lines are by definition perpendicular to equipotentials and are therefore pulled into the sphere as shown in Figure 3.4. The field lines indicate the paths along which free charge can travel so it can be seen that the sphere will acquire charge.

Field lines between a high-voltage point and a plane are shown in

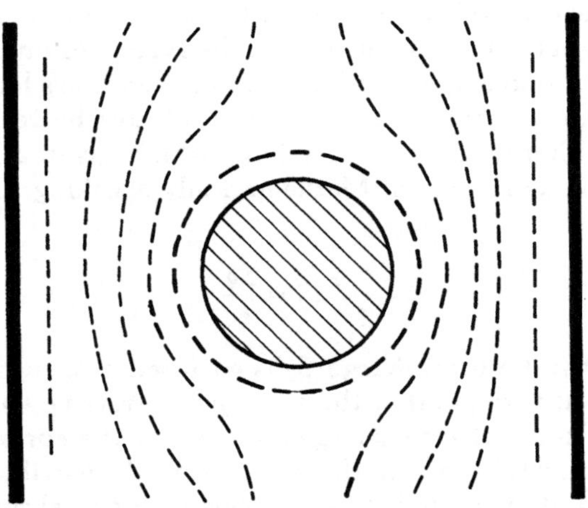

FIGURE 3.6. Equipotentials around a conducting sphere between two plane electrodes.

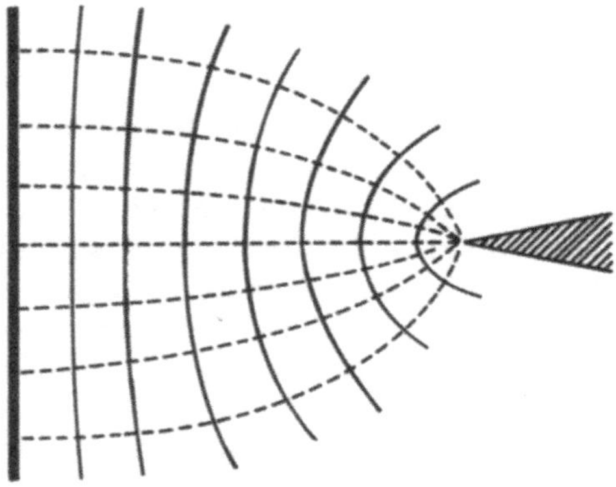

FIGURE 3.7. Field intensification at a point.

Figure 3.7. The very high field at the point is clearly shown by closely spaced lines. If the field exceeds 3×10^6 V/m the air ionizes and there is a corona discharge.

3.2.3.2. Coulomb's Law and Gauss's Law Relating Charge and Field

A charge q at a point creates an electric field E given by the equation

$$E = \frac{q}{4\pi\epsilon_0 r^2} \tag{3.5}$$

where q is in coulombs and r in meters; the field is directly proportional to the charge and inversely proportioned to the square of the distance from it. The constants of proportionality fix the system of units chosen. This law was first found empirically by Coulomb and bears his name.

In practice a charged system cannot always be represented by a single point charge and Gauss's Law which gives a more general formulation is more useful. Mathematically speaking, Gauss's Law is written

$$\int E\,dS = \frac{\Sigma q}{\epsilon_0}. \tag{3.6}$$

This states that if a total charge Σq is enclosed by a surface S then the electric field* at each point of the surface integrated over the surface is proportional to the electric charge enclosed by the surface. Where the surface is a simple shape with the field uniformly distributed over it the field at any point may simply be multiplied by the total surface area. For

*More precisely, the component of the field perpendicular to the surface.

example, the electric field E at the outside of a spherical cloud of charged powder of radius r and charge density Q C/m³ is given by

$$E(4\pi r^2) = Q\frac{4\pi r^3}{3\epsilon_0}$$

i.e.,

$$E = \frac{Qr}{3\epsilon_0} \tag{3.7}$$

Thus the charge density in a cloud can be found from a measurement of electric field on the outside.

3.2.4. Resistivity

The electrical resistance R (measured in ohms, Ω) is defined as the ratio V/I, where I is the current flowing when a voltage V is applied. The resistance of any path increases as its length increases and decreases as its area increases. Resistance is therefore more conveniently expressed in terms of resistivity ρ, i.e., the resistance of unit length of a material of unit area.

$$\rho = \frac{RA}{d} \tag{3.8}$$

ρ is expressed in $\Omega \cdot$ m. The resistivity is a fundamental property of the material and does not depend on the dimensions. The resistivity is by definition the reciprocal of conductivity σ

$$\rho = \frac{1}{\sigma} \tag{3.9}$$

The most important use of a measurement of resistivity is to give an estimate of the time taken for any charge created to leak away. The time constant τ is given by $\tau = \epsilon\epsilon_0\rho$, where ϵ is the relative dielectric constant and ϵ_0 is the permittivity of space, 8.8×10^{-12} F/m. The dielectric constant changes by a factor of only a few times, while resistivity may change by many orders of magnitude. Hence it is the resistivity that dominates the charge decay and it is usually adequate to consider $\epsilon\epsilon_0$ as approximately equal to $2 = 10^{-11}$.

It is important to distinguish between bulk and surface resistivity. Many insulators conduct rather better along the surface than through the bulk of the material, and it is then more relevant to consider the resistance along a surface than the bulk resistivity. If measurements of surface resistance are made between electrodes whose spacing equals their length then the dimension chosen for length and spacing does not affect the measurement. One method of measuring surface resistance is to apply electrodes as two sides of a square. The resistance is then quoted in ohms/square (Ω/□) or simply in ohms (Ω).

For many insulating materials and for powders the resistivity is a function of the voltage at which it is measured. The familiar relationship by which current and voltage are related by a constant resistance (Ohms' Law) cannot be assumed to apply to these materials.

3.2.5. Capacitance

The capacitance of an object is defined as the ratio between the charge Q stored on the object and V, the rise in potential this charge causes:

$$C = \frac{Q}{V} \tag{3.10}$$

The capacitance of particular configurations can be calculated, with examples given in most textbooks on electricity and magnetism (e.g., Bleaney and Bleaney, 1957). The results of two such calculations are particularly useful because they represent a reasonable approximation to many real situations where an order of magnitude calculation of capacitance is needed.

A parallel-plate capacitor (or two parallel charged surfaces) of area A separated by a thickness d of a dielectric with permittivity ϵ has a capacitance C given by

$$C = \frac{\epsilon\epsilon_0 A}{d} \tag{3.11}$$

The capacitance between the surface of a conveyor belt and a roller, and other similar nonplanar situations when d is small, can be reasonably represented by this formula, which is also useful in estimating the capacitance between field meters and the charged surface being measured.

An isolated sphere of radius a in air has an intrinsic capacitance of

$$C = 4\pi\epsilon_0 a \tag{3.12}$$

and this equation can be used to give a rough order of magnitude of the intrinsic capacitance of any large object. For example if a is 1 m, C is ~ 110 pF. The capacitance of a man is between 100 and 300 pF.

3.2.6. Poisson's Equation

The relationship between electric field and charge in the general case is given by Gauss's theorem.

$$\int E\,dS = \frac{\Sigma q}{\epsilon_0}$$

This can also be written in a different mathematical form.

$$\text{div } E = \frac{Q}{\epsilon\epsilon_0} \tag{3.13}$$

where Q is the charge per unit volume.

E is by definition the rate of change of potential, i.e.,

$$-\text{grad } V = E$$

therefore

$$\text{div(grad } V) = -\frac{Q}{\epsilon\epsilon_0}$$

This can be written

$$\nabla^2 V = -\frac{Q}{\epsilon\epsilon_0} \tag{3.14}$$

and is known as Poisson's equation. In principle, this equation allows the potential due to a known set of charges and conductors to be calculated at any point. Any problem involving electrostatic fields may be solved by finding a solution that satisfies Poisson's equation and fits the boundary conditions for field and potential. Generally, these conditions are the potentials applied to conductors and the fact that there can be no field along the surface of a conductor. For a simple geometry the equation can be solved analytically, and standard solutions such as the calculation of the field distortion for a sphere in a uniform field can be found in most textbooks on electromagnetism. The equation can also be used to map potentials in a complicated situation such as in the inside of a powder-storage silo. In this case the solution is found by computer using finite element methods.

A computer program is set up to divide a three-dimensional vessel into a mesh of three-dimensional elements. Known charge densities and fixed potentials are set and Poisson's equation solved for each element. Standard programs have been written so that the user need only define the geometry of the system and the position of regions of different charge densities to produce a potential map of the whole system. In some cases the charge density is not known accurately and the program can then be used to try out a different configuration and see which best fits measurements made with field meters. Use of a computer program to look at a powder-storage silo allows a much more-detailed solution of potentials to be defined than could be achieved by measurement alone.

Interpretation of field-meter data in a powder cloud is difficult without a program that can take account of nonuniform charge distributions. The computer solution can also make allowance for the changes in field brought about by the meter itself and, from the measurements taken, give data on the electric field and potential in the absence of a meter.

3.2.7. Discharging of Capacitors

An electrostatic spark occurs when an isolated object which has been allowed to accumulate charge is suddenly grounded. An equivalent electrical circuit can be drawn for this (Figure 3.8). The charged object is a capacitor. The gap across which the spark jumps has some resistance to current even when the spark is present so this must be represented by a resistance R. When the electrode A is moved to position A′ the gap breaks down and the capacitor discharges. It does not lose its charge instantaneously because of the resistance of the gap, but the charge in the capacitor decays exponentially with time constant τ

$$Q = Q_0 \, e^{-t/\tau} \tag{3.15}$$

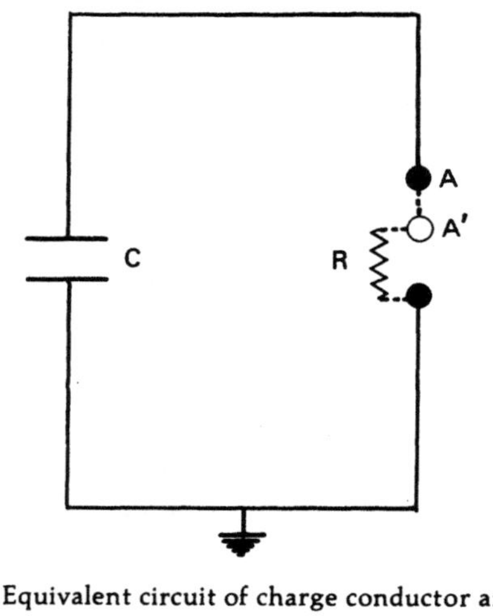

FIGURE 3.8. Equivalent circuit of charge conductor and spark gap.

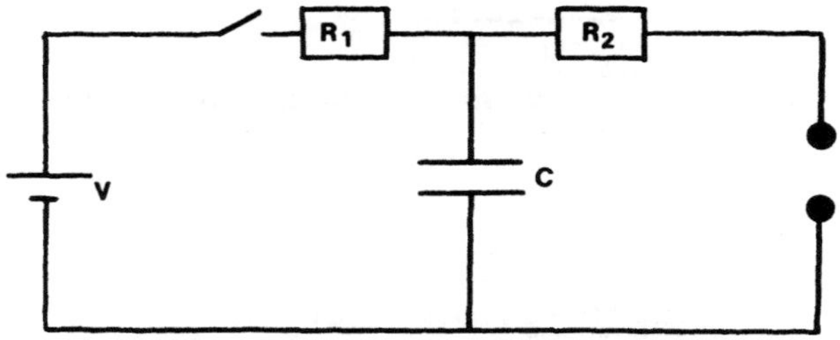

FIGURE 3.9. Circuit to simulate spark discharge.

τ is known as the *time constant* and is equal to the product RC. In practice if there is no additional series resistance between the capacitor and ground when contact is made (for example, a human being would add a resistance of $\sim 10^5–10^7\ \Omega$), then the resistance of the gap is a few hundred ohms and the spark usually lasts a few nanoseconds. Sparks from human beings or liquid surfaces are longer in duration—up to a few tens of microseconds. Electrostatic sparks may be simulated in the laboratory using the circuit shown in (Figure 3.9). The capacitor C charges through resistor R_1. The charge in the capacitor increases exponentially with a time constant R_1C. The discharge through the spark gap has a time constant R_2C assuming R_2 is much greater than the gap resistance. The capacitor voltage can increase up to the voltage V. Therefore the energy stored is $\frac{1}{2}CV^2$. Some of this energy remains in the capacitor on discharge and some is lost in the resistor R_2. Some circuits replace R_2 with an inductor (L) which increases the spark duration but does not absorb energy.

3.3. MEASUREMENT TECHNIQUES

3.3.1. Electric Field

Although electric charge is the most fundamental quantity of electrostatics, it cannot always be measured directly in a practical situation. Potential is also difficult to measure because it is altered by a change in capacitance and most instruments affect the capacitance of the system they are measuring. The electric field is therefore the most universal measurement used to calculate both charge and potential.

3.3.1.1. *Induction Field Meter*

The simplest type of field meter is an induction probe. An electric field induces a charge on the detecting plate of the instrument (Figure

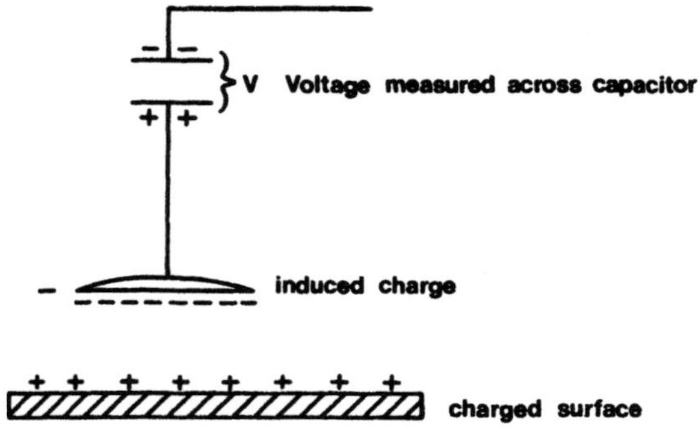

FIGURE 3.10. Induction field meter.

3.10). The meter reads the voltage that this charge generates across the internal capacitance of the instrument. The induced charge is proportional to the electric field and hence gives a measure of it. The induction field meter is very simple in concept. In practice it tends to suffer from zero drift as a result of signals generated by the bias current requirements of the amplifier. Atmospheric ion currents or free charge from the collection of charged powder particles or from discharges also charge the meter capacitance and give an error signal. The meter is inexpensive and can be satisfactory where measurements are needed only a short time after setting zero. It cannot be used for continuous monitoring or in an environment with free ions or charged particles.

3.3.1.2. *Rotating-Vane Field Mill*

The most widely used instrument for monitoring electric fields is the field mill. In principle it is an induction field meter with a grounded butterfly rotor placed in front of it (Figure 3.11). As the rotor turns the sensor plate is alternatively exposed and shielded from the field and an alternating voltage is generated across the meter capacitance. The peak voltage is proportional to the electric field, and independent of the rotor speed. The sensitivity of the instrument is changed by adjusting the value of the internal capacitance. A phase-sensitive detector using the shutter rotation as a reference frequency extracts the field information. The amount of free charge able to reach the sensor plate in a single cycle ($<1/100$ s) is negligible and this instrument can therefore be used in the presence of free charge.

There are a wide variety of field mill instruments commercially available. In the majority a small electric motor is used to drive the rotor, but for hazardous environments clockwork or air turbine drives are

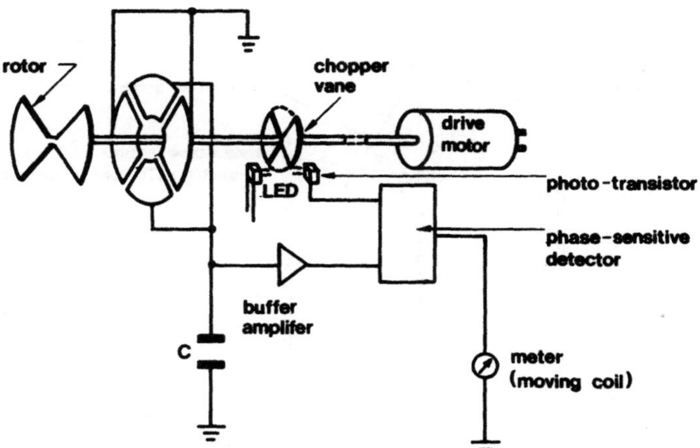

FIGURE 3.11. Rotating vane field meter.

available. To satisfy safety requirements, circuit modifications are also made so that it is impossible for incendiary discharges to occur in the measuring system even in the event of component failure.

3.3.1.3. *Vibrating Grid Field Meter*

A similar principle is used in the vibrating grid field meter. In this instrument an oscillating grid or aperture is placed in front of the sensor plate. The extent to which the field penetrates to the sensor plate and the extent of the charge induced on it depend on the position of the grid. Therefore, an oscillatory signal is produced on the plate, from which the electric field can be found.

3.3.1.4. *Radioactive Field Meter*

A completely different principle is used in the radioactive field meter (Figure 3.12). A radioactive source ionizes the air within the meter so that the path between the meter and the charged surface being measured and the path between the radioactive source and the ground are weakly conducting. These two paths form a potential divider and the voltage at the source therefore assumes a value that is a fraction of the surface voltage. The value of this voltage will depend on the ratio of the resistance of the two paths, and hence on the relative separation between the radioactive source and the unknown charge and the source and ground.

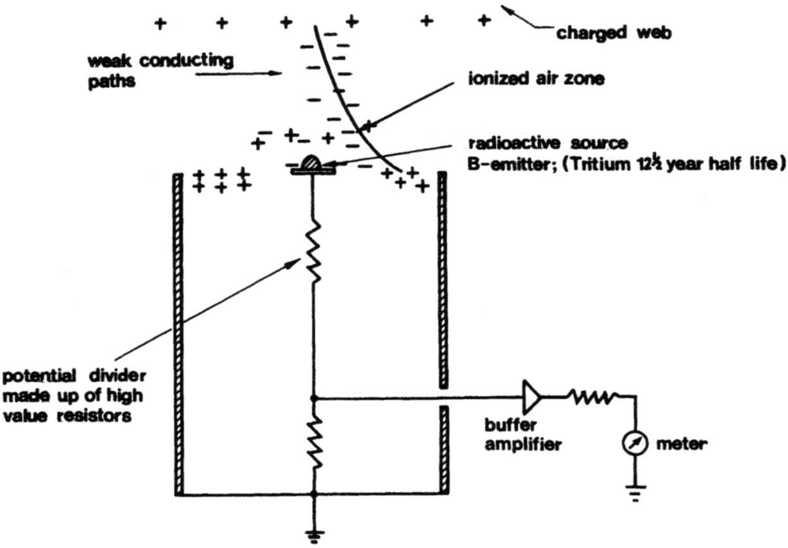

FIGURE 3.12. Radioactive field meter.

The potential at the source is measured using the internal potential divider and buffer amplifier. For this instrument the meter reading changes as the spacing of the instrument is changed. Since usually the charge density is invariant and the measured potential changes with capacitance, the positioning of the meter affects both the sensitivity of the meter and the actual potential being measured. The radioactive meter must therefore be operated at fixed distances for which its output has been calibrated. It then gives the voltage on the surface. This voltage must be divided by the spacing to convert to the electric field and then the surface charge densities can be calculated.

The sensitivity of the instrument is adjusted by changing the spacing as directed and also by changing the size of the aperture through which the external field enters the ion chamber. The radioactive source used is polonium which is an α emitter. Alpha particles are absorbed in only a few centimeters of air and therefore present no radiation hazard. The polonium-210 source has a half-life of 138.4 days and must therefore be replaced regularly.

3.3.1.5. Comparison of Different Types of Field Meter

The radioactive field meter has a longer response time than the rotating vane instrument and requires several seconds to settle to a new field level. However, at present it is the cheapest of the intrinsically safe meters on the market. The biggest problem with radioactive field meters is the psychological disadvantage they possess of having a radioactive source at all. Although the source is completely safe in this instrument application, the unjustified suspicion it generates in operators and other personnel is often cited as a reason for choosing one of the other types of monitoring instruments. The need for source replacement and the very slight possibility of polonium contamination of food and pharmaceutical products must, however, be considered.

The induction-type instrument is useful as a hand-held device provided there is a low level of free charge but frequent rezeroing is a necessity.

Both the rotating vane and the radioactive devices are sold in an intrinsically safe form with BSEEFA and PTB certification for use in flammable-vapor environments. Neither is specifically designed to be dustproof and generally field meters are purged with an air jet to keep them free from powder when used in a highly dusty environment.

The field meter is probably the most versatile electrostatic instrument and the most difficult to interpret. The biggest problem is that it is a fairly large grounded object that is inserted into the system. Unless it is situated in a grounded position (a place of zero potential) it will distort the field it is measuring. In simple geometries, such as an isolated sheet of charge, the effect of the meter on the field to be measured can be calculated, but in more-complicated geometries, such as the inside of a powder-storage silo, the true field must be found from the measured field at various positions using a computer program.

3.3.2. Use of a Field Meter for Measurement of Charge and Voltage

3.3.2.1. *Voltage*

All field meters basically measure the electric field in volts/meter or volts/foot according to the calibration. However, since field, charge, and voltage bear simple relationships to each other the instrument can also be used to measure charge and voltage. In a parallel-plane geometry the electric field is uniform and is the potential difference divided by the spacing, i.e., $E = V/d$. A field meter placed a distance d from a plate carrying a potential V will measure a field E. Thus, the potential is simply found by multiplying the field in volts/meter by the spacing in meters. If the object carrying the voltage is not a plane the field is not uniform and it is not strictly accurate to assume $V = Ed$, however, for many purposes this gives a reasonable approximation.

Some commercial field meter manufacturers assume that the

instrument will be used mainly for potential measurement and calibrate the meter in volts stating the distance at which it must be held. This practice is rather confusing, particularly when it is more convenient to use other than the specified spacing.

3.3.2.2. *Surface Charge*

A uniformly charged surface generates an electric field, perpendicular to the surface, that is proportional to the surface charge density. The relationship between surface charge and field is found from Gauss's Law, equation (3.6). Gauss's Law is applied to a cylinder imagined to intersect the charged surface as shown in Figure 3.13. For a charged metal surface with surface density σ, the field lines will be perpendicular to the surface and cannot exist in the metal. Therefore E, the field measured in the air above the surface, is given by

$$\int E\, dS = \frac{\sigma dS}{\epsilon_0}$$

or

$$\sigma = \epsilon_0 E$$

This applies only to a *charged conductor*. For a charged insulator a field may also exist in the material and although the field will still be proportional to the charge, the constant of proportionality may not be ϵ_0. A field meter can therefore be calibrated to measure surface charge. However, interpretation is considerably complicated because the meter adds capacitance to the system. The charge to be measured is shared between the two capacitances represented by the charged surface and ground and the charged surface and the field meter. These capacitors are effectively in parallel as the meter itself is a virtual ground. The relative sizes of these capacitances must be estimated to obtain the true charge.

If the charged surface is well isolated and the field meter is held very close to the surface, then the meter-to-surface capacitance may be

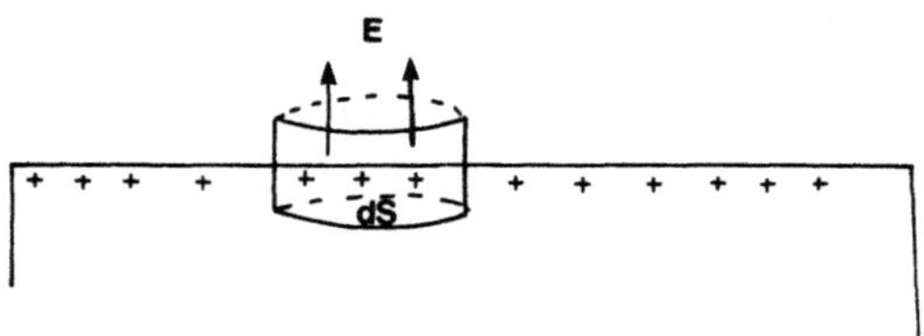

FIGURE 3.13. Field due to a surface charge.

assumed to dominate. If, on the other hand, the charge is to be measured on a thin film passing over a metal roller, the film-roller capacitance dominates. This can be calculated using the relationship [equation (3.11)]

$$C = \frac{\epsilon\epsilon_0 A}{d}$$

The charge can be found from the measured surface potential V using

$$C = \frac{Q}{V}$$

An example of this measurement is given in detail in Section 3.4.

3.3.2.3. *Volume Charge Density*

The charge density in a cloud of powder can also be found from a field meter measurement by calculation from Gauss's Law which relates electric field to the charge that causes it

$$\int E\, dS = \frac{\Sigma q}{\epsilon_0}$$

Σq is the total charge enclosed by the surface. In an unconfined cloud Σq and S are not easily defined; therefore, in practice a field meter is set up in one wall of the grounded mesh container. Ideally the charged powder is able to pass freely into the container through the mesh which acts as a screen between the meter and all external charge. The mesh provides the total surface S and the field E at the mesh is measured by the field meter. The product of the field E and the total surface area of the mesh multiplied by the constant ϵ_0 is equal to the total charge enclosed. It is probable that at high charge levels the charge on the dust will be somewhat reduced as it enters the screen.

3.3.3. Electric Potential

Electric potentials in electrostatics are measured in two circumstances: to evaluate the hazard of static accumulation and to test the high-voltage supplies of electrostatic equipment such as electrostatic paint spray systems, electrically powered antistatic bars, and ion-generating devices. Different types of instrument are often used in the two applications because it is possible to draw a small current from a power supply without reducing the voltage whereas a true electrostatic potential is reduced if some of the charge creating it flows to ground through a measuring instrument.

The potential of a power supply may therefore be measured either by allowing a current to flow through a resistor or electrostatically. An electrostatic potential on the other hand must be measured by an instrument that draws no current and preferably has only a small electrical capacitance to ensure that charge transfer is small.

3.3.3.1. *Measurement of the Voltage of Power Supplies*

High-voltage power supplies usually have a very high output impedance to prevent serious injury in the event of accidental contact by an operator. Therefore, the measuring instrument must have a very high input impedance to ensure that the voltage being measured is not affected.

The operation of two resistor-based voltage measuring devices is illustrated in Figure 3.14. In diagram (a) the current flowing through the high resistance is measured and the voltage is found using Ohms' Law.

$$V = IR \tag{3.16}$$

In diagram (b) the voltage is divided between the high resistance R_1 and the low resistance R_2. A conventional voltmeter can then be used to measure the low voltage V_2 across resistor R_2.

$$V_1 = \frac{V_2(R_1 + R_2)}{R_2} \tag{3.17}$$

The resistance value required is normally around $10^{11}\ \Omega$ so that only 1 μA is drawn for 100 kV. It is important to consider the possibility of errors

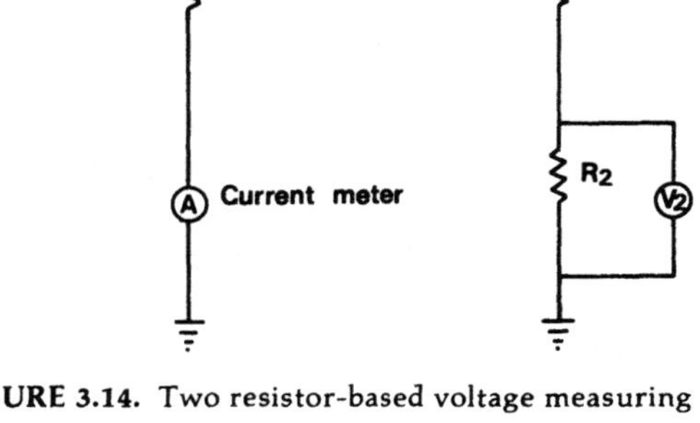

FIGURE 3.14. Two resistor-based voltage measuring devices.

from conduction across dirty resistors and inadequate insulation and also the risk of corona from highly stressed regions, which must be avoided by eliminating sharp edges. Air insulation is possible for voltages up to 100 kV, but above this level oil emersion or encapsulation in epoxy or silicon rubber is necessary.

A similar technique can be used for AC measurements but the resistors used to divide the voltage must be replaced with a resistor/capacitor combination. Low-frequency AC signals where the root-mean-square voltage does not vary rapidly can be measured most simply with an electrostatic voltmeter (discussed in the next section), but transients must be measured using a high-voltage divider in combination with an oscilloscope. Commercial dividers which work at high frequencies seldom have a sufficiently high impedance for electrostatic applications, and the measurement of rapidly changing high voltages, such as the potential during a spark, is extremely difficult. The measurement of transient high voltage is discussed by Schwab (1972).

3.3.3.2. *Electrostatic Voltmeter*

Electrostatic voltmeters make use of the force exerted between two objects at the same potential. Typically, the instrument has two interleaving sets of vanes. One set is fixed and the other mobile and attached to a pointer of some type. On application of a voltage the moving vanes are displaced by a deflection force proportional to the square of the potential and the pointer moves across a dial. No current is drawn other than that needed to charge the capacitance represented by the vanes. The input resistance can theoretically be infinite although in practice it is limited by the resistance of mountings. The deflection is usually measured directly by detecting the movement of a beam of light reflected from the moving parts. Since the force is proportional to the square of the voltage the scale is nonlinear and rather cramped at the low-voltage end of the scale, but this disadvantage is offset by the ability to read root-mean-square values of alternating potentials and by the very high input resistance. The capacitance of these meters is usually between 2 and 20 pF. If an electrostatic voltmeter is to be used to measure the potential on an object whose capacitance is not at least an order of magnitude greater than the meter capacitance an adjustment must be made for charge sharing. An example is given in Section 3.4.

3.3.3.3. *Field Meters*

The use of various types of field meter to measure potential was discussed in Section 3.3. The field meter does not contact the surface for which it is measuring the voltage; therefore, it leaks no charge. In other words, it effectively has an infinite input resistance. The input capacitance is simply the stray capacitance between the surface being measured and the meter and hence is very small. Charge sharing is therefore small. The two disadvantages of using a field meter for voltage measurement are the need for estimating the stray capacitance and the requirement for parallel-plane geometry. Most field meters will measure only DC potentials but, recently, field mill voltmeter operation has been extended up to 50 Hz. This has been achieved by increasing the number of rotor/sensor plate segments to give a higher internal operating frequency (~2 kHz). The instrument effectively monitors the instantaneous voltage at a series of points on each half-cycle to reproduce the AC signal at the output of the phase-sensitive detector.

The field meter is the only method available for measuring electrostatic potentials on insulating surfaces because both electrostatic voltmeters and potential dividers require some movement of charge. They are also necessary for moving surfaces such as conveyor belts and webs where no direct contact can be made. The field meter is used indirectly to measure potential in the field meter voltmeter. The potential to be measured is connected to a sphere and a field meter at a fixed distance from the sphere measures the voltage on it. The nonlinearity of the geometry introduces no inaccuracy because the field meter/voltmeter can be directly calibrated. No current is drawn by this instrument if the sphere is well isolated but there is some charge transfer to the sphere which leads to inaccuracy if the voltage is stored on a low-capacitance object.

3.3.3.4. Spark Gaps

It was found by Paschen in 1889 that the sparking potential of a gap depended on the air density and the gap length only. It is therefore possible to estimate potentials by measuring the minimum gap between the test sphere and voltage source which can be achieved before a spark occurs. The breakdown potential for that gap can then be found from standard tables. It is found experimentally that there is a random variation in the results obtained, so this method gives only approximate voltages unless several measurements are made. The scatter is reduced considerably in the presence of an ionizing point of UV radiation, which helps initiate the spark. This is also found to reduce the breakdown potential by approximately 3.5%.

Variations can be reduced by standardizing the system. The high voltage is connected to a sphere in a standard container and the second sphere is brought up to produce a spark. There must be no insulation

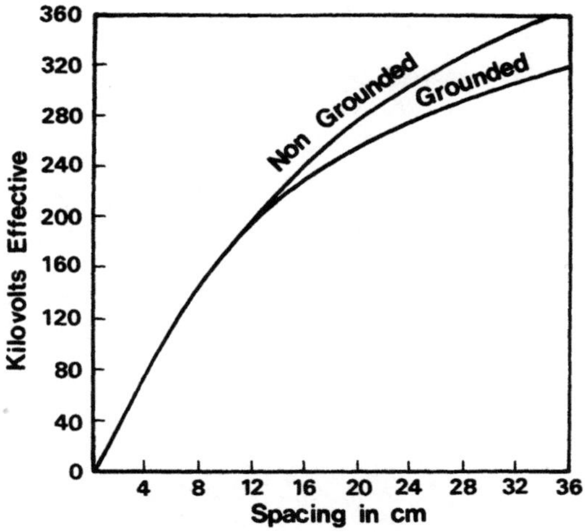

FIGURE 3.15. Breakdown voltage of sphere gap.

closer than one sphere diameter to the spark gap. The breakdown voltage will differ according to whether one sphere is grounded or not. This is shown in Figure 3.15. Spark gaps are used mostly for measuring very high voltages above 100 kV but can provide a useful guide at lower voltages. The breakdown potentials for different sphere diameters and spacings are given in Table 3.1.

3.3.4. Charge

The charge is the most important parameter in electrostatics because without charge no potential is generated and no field exists. It is the fundamental parameter and the total charge is not changed by instrumentation. It is created by the processes discussed earlier and lost either by discharge (usually breakdown of the air) or by leakage at a rate fixed by the resistance of the available path.

3.3.4.1. The Faraday Cup

Charge can be measured directly by a Faraday cup (or *can* as it is sometimes called), provided the charge is stored on something that can be placed in a container. The Faraday cup is a double-walled vessel illustrated in Figure 3.16. The outer wall is grounded and provides a screen to prevent stray external charges from affecting the measurement. If a charged object is placed into the vessel, an equal and opposite charge is induced on the inner wall. This charge is measured directly with an

TABLE 3.1. Flashover Voltages: One Sphere Grounded[a]

Sphere-gap spacing	Sphere diameter, cm											
	2	5	6.25	10	12.5	15	25	50	75	100	150	200
0.05	2.8											
0.10	4.7											
0.15	6.4											
0.20	8.0	8.0										
0.25	9.6	9.6										
0.30	11.2	11.2										
0.40	14.4	14.3	14.2									
0.50	17.4	17.4	17.2	16.8	16.8	16.8						
0.60	20.4	20.4	20.2	19.9	19.9	19.9						
0.70	23.2	23.4	23.2	23.0	23.0	23.0						
0.80	25.8	26.3	26.2	26.0	26.0	26.0						
0.90	28.3	29.2	29.1	28.9	28.9	28.9						
1.0	30.7	32.0	31.9	31.7	31.7	31.7	31.7					
1.2	(35.1)	37.6	37.5	37.4	37.4	37.4	37.4					
1.4	(38.5)	42.9	42.9	42.9	42.9	42.9	42.9					
1.5	(40.0)	45.5	45.5	45.5	45.5	45.5	45.5					
1.6		48.1	48.1	48.1	48.1	48.1	48.1					
1.8		53.0	53.5	53.5	53.5	53.5	53.5					
2.0		57.5	58.5	59.0	59.0	59.0	59.0	59.0	59.0			
2.2		61.5	63.0	64.5	64.5	64.5	64.5	64.5	64.5			
2.4		65.5	67.5	69.5	70.0	70.0	70.0	70.0	70.0			
2.6		(69.0)	72.0	74.5	75.0	75.5	75.5	75.5	75.5			
2.8		(72.5)	76.0	79.5	79.5	80.0	81.0	81.0	81.0			
3.0		(75.5)	79.5	84.0	85.0	85.5	86.0	86.0	86.0	86.0		
3.5		(82.5)	(87.5)	95.0	97.0	98.0	99.0	99.0	99.0	99.0		

4.0	(88.5)	(95.0)	105	108	110	112	112	112	112		
4.5		(101)	115	119	122	125	125	125	125		
5.0		(107)	123	129	133	137	138	138	138	138	
5.5			(131)	138	143	149	151	151	151	151	
6.0			(138)	146	152	161	164	164	164	164	
6.5			(144)	(154)	161	173	177	177	177	177	
7.0			(150)	(161)	169	184	189	190	190	190	
7.5			(155)	(168)	177	195	202	203	203	203	
8.0				(174)	(185)	206	214	215	215	215	
9.0				(185)	(198)	226	239	240	241	241	
10.0				(195)	(209)	244	263	265	266	266	266
11.0					(219)	261	286	290	292	292	292
12.0					(229)	275	309	315	318	318	318
13.0						(289)	331	339	342	342	342
14.0						(302)	353	363	366	366	366
15.0						(314)	373	387	390	390	390
16.0						(326)	392	410	414	414	414
17.0						(337)	411	432	438	438	438
18.0						(347)	429	452	462	462	662
19.0						(357)	445	473	486	486	486
20.0						(366)	460	492	510	510	510
22.0							489	530	555	560	560
24.0							515	565	595	610	610
26.0							(540)	600	635	655	660
28.0							(565)	635	675	700	705

[a] Kilovolts peak at 20°C, 1013 mbars.

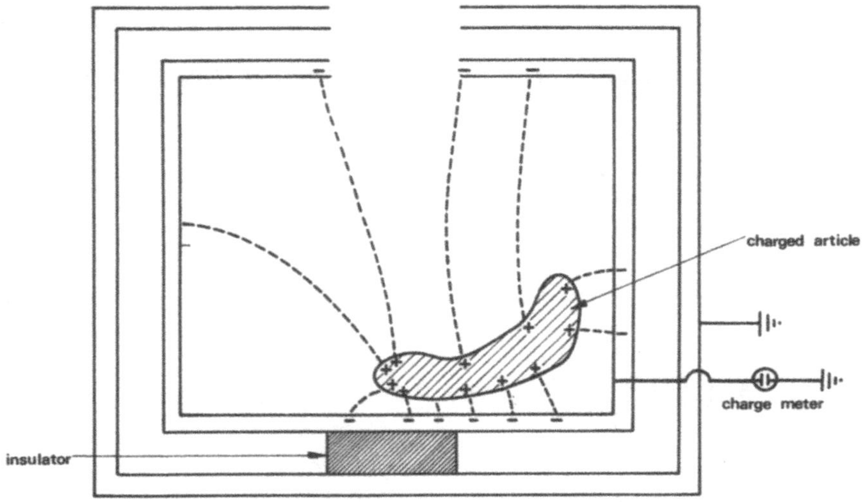

FIGURE 3.16. Faraday cup.

electrometer. Charge does not necessarily have to flow from the test article to the meter since an equal and opposite charge flows from the meter to the can to balance the unknown charge inside.

The electrometer has an internal capacitor which becomes charged by the flow of charge to the can. Hence, if the electrometer is zeroed when the article is in place, the meter capacitance is discharged. Moving the charged article from the can then leaves the image charge on the can wall unbalanced and the meter will recharge but read the opposite polarity. It is necessary to take care to reset to zero before each measurement. Insulation between the inner and outer can must be of high resistance to avoid leakage.

The Faraday cup cannot be used when the charge is on a moving belt or a large article, but it is useful for estimating the charge on a liquid or powder sample. Powder charge levels are normally quoted in coulombs per kilogram (C/kg). Care must be taken that the means by which powder or liquid is introduced into the cup does not produce additional charge. The powder must be caught directly in the can and must not be transferred indirectly or passed through a tube or funnel. When the powder is suspended in an air stream the inner can is provided with holes to allow the air to escape and is lined with cloth or porous paper. Obviously high-speed pneumatically conveyed powders cannot be practically trapped by this technique.

3.3.4.2. Charge on Powders

The Faraday cup or can measures the total charge entering the inner container. Therefore if a powder consisting of both positive and negative particles is to be measured, only the net charge will be detected. The charge on an individual particle can be measured by allowing it to fall under gravity in a transverse electric field. The particle is photographed while being illuminated by a flashing light and the path is then reproduced on the film. Analysis of the distance traveled vertically and horizontally gives the relative gravitational and electric force. From this the charge-to-mass ratio of the individual particle can be calculated. Deflection in an electric field has also been used to separate positive and negative particles from a mass of flowing powder but this has not been found to be a very accurate measuring technique.

3.3.5. Field and Potential Probes

The potential at a point in space may be measured by means of a fine-wire probe. This is introduced as far as possible along an equipotential line in order to not disturb the situation. Its potential is allowed to gradually change to the potential that exists along that line. The end of the wire is sharply pointed so that it can create a corona discharge at its tip, giving off sufficient free charge to allow the potential to change. At present the technique is not commonly used outside the laboratory because the probes are not sturdy, but further development could make this technique very useful for investigating potentials inside powder-storage vessels and large containers.

Probes may also be inserted into pipe lines, but here the measurement is distorted by the frictional charge produced by the powder moving rapidly against the probe. Although there is scope for development of probes to be used in clouds and pneumatic lines, at present the field meter is the only satisfactory instrument. Of course, even this must be carefully purged to prevent powder from collecting in the instrument and damaging it.

A novel type of probe proposed by Cooperman (1956) has been tested in powder clouds by Corbett (1972). A small metal ball is fired through the region to be measured. The charge it collects is given by Pauthenier's Law, which was discussed in Section 3.2.1.2., and is proportional to the electric field. Again, ions must be present to enable the ball to charge but the powder itself becomes strongly charged if sufficient ions are present for the probe to operate. Therefore, this is only of use as a laboratory tool and in highly charged powder clouds.

3.3.6. Resistivity

3.3.6.1. Volume Resistivity of a Bulk Solid

The resistivity of a material is defined as the electrical resistance of unit length of a material of unit area. It is therefore essentially a measurement of a block of solid and cannot be applied to a surface. The volume resistivity is found by measuring the current flowing between electrodes of area A and spacing d when a known voltage V is applied. The resistance R of the sample is given by

$$R = \frac{V}{I}$$

In a geometry with plane parallel electrodes, ρ is given by

$$\rho = \frac{RA}{d}$$

A/d is known as the *cell constant*. Resistivity is a fundamental property of the material and not dependent on cell geometry, whereas the resistance is a function of the sample size.

Liquids and powders are sometimes measured in a cell where electrodes are in the form of concentric cylinders. The cell constant (usually supplied by the manufacturers) is then dependent on the ratio of the radii of the two cylinders.

In a solid sample the surface is often more conducting then the bulk. Conduction between the electrodes on opposite sides of the solid, via the surface, must then be prevented by guard rings. Each electrode is surrounded by a ring at the same potential but powered from a different supply. Only the current flowing between the two central electrodes is measured. No current flows across the surface from the central electrodes because there is no potential difference between the ring and the electrode. The guard ring also helps to make the field across the bulk material uniform as shown in Figure 3.17.

Insulating materials may have resistivities from 10^{14} to 10^{18} $\Omega \cdot$ m. Measurements therefore require application of high voltages and detection of very low currents. With solid materials electrodes are often painted onto the surface with conducting paint to ensure that a good contact is made with the surface. The resistivity measured in this way is often an order of magnitude lower than that measured with solid electrodes pressed to the surface even when the applied pressure is quite high.

When the material is highly insulating the number of charge carriers is small. The application of a high voltage to a test electrode can inject

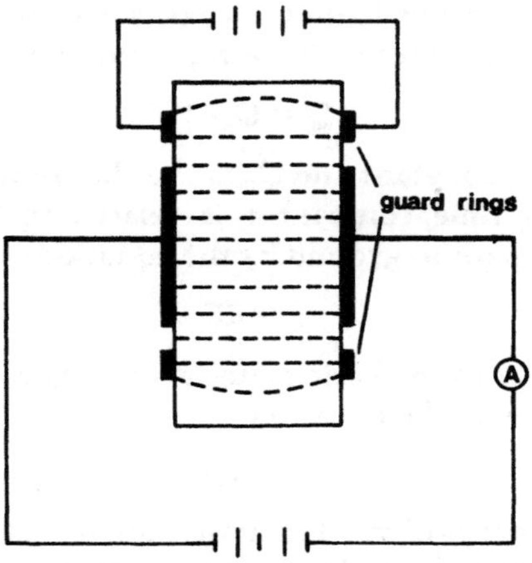

FIGURE 3.17. Measurement of bulk resistivity.

charge carriers so that a higher conductivity is measured than really exists. This can be avoided by the four-electrode technique illustrated in Figure 3.18. A current I is passed through a slab of material by application of a voltage. The voltage V is then measured between two other points on the surface. The resistance of the material between the two voltage test points is then V/I. Again, guard electrodes must be used to prevent surface leakage.

In any measurement of a high-resistance material there will be a high initial current flow that gradually decays to a steady level. The high initial current is the displacement current and represents the energy that must be absorbed to polarize the atoms of the insulator. The final steady current is the leakage current from which the resistivity is found.

The importance of resistivity in the evaluation of electrostatic hazards is primarily due to the direct relationship between the time required for the charge to decay and the resistivity.

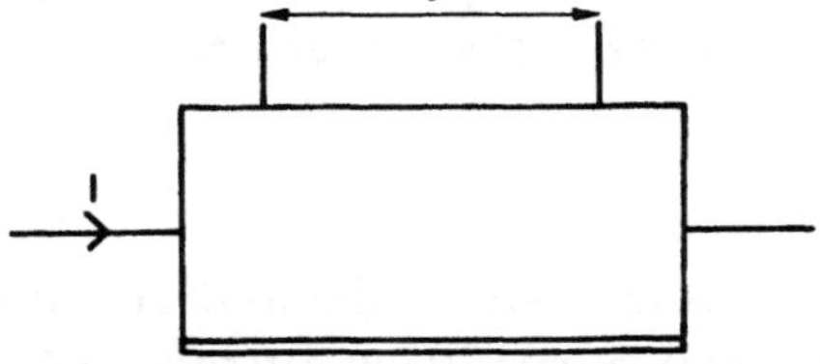

FIGURE 3.18. Four-electrode technique for measuring resistance.

If Ohms' law is obeyed and the charge decays through a constant resistance R, then the charge will decrease exponentially

$$Q = Q_0 e^{-t/\tau} \tag{3.18}$$

where τ is the time constant and Q_0 is the charge when the time $t = 0$. The charge-decay time constant τ is related to the resistance and capacitance of the path to ground by the equation

$$\tau = RC$$

τ can also be related to the basic material properties of resistivity ρ and dielectric constant ϵ by the equation

$$\tau = \epsilon\epsilon_0 \rho \tag{3.19}$$

The dielectric constants of most materials other than metals lie between 1 and 10, whereas the resistivities vary over many orders of magnitude. The resistivity therefore affects the charge-decay time (and hence the extent to which charge is stored) more than the dielectric constant.

3.3.6.2. *Surface Resistivity of Bulk Solid*

The resistivity of the surface of a solid is measured between parallel electrodes placed on the surface in such a way that the length of the electrode is equal to the spacing between them. A voltage is applied between the electrodes and the current measured using an ammeter or electrometer. Provided the electrode spacing and length remian equal, the actual dimensions chosen do not affect the value of surface resistivity and the same numerical value is obtained in ohms for a square centimeter, a square meter, or a square foot. Surface resistivity may therefore be quoted in ohms/square (Ω/□) to indicate the method by which it is measured.

Surface resistivity is sometimes expressed in Ω · m by multiplying by the area of the contacting electrodes and dividing by their spacing, as in the measurement of bulk resitivities. It can be easily seen that in a true two-dimensional measurement the area of the electrode should not affect the measurement—only the length. Hence this method of expressing surface resistivities is not realistic and should not be used.

3.3.6.3. *Charge Decay*

It is not at all simple to relate the theoretical charge-decay time to the surface resistance and it is best to measure decay time directly, using a field matter. The field above the surface is proportional to the charge on the surface. A field meter thus monitors the rate of decay of charge. If the charge decay were truly resistive it would be exponential, i.e.,

$$Q = Q_0 e^{-t/\tau}$$

where $\tau = RC = \epsilon\epsilon_0\rho$.

Thus

$$\log Q = -\frac{t}{\tau} Q_0$$

The logarithm of Q plotted against time should give a straight line with a slope that is the reciprocal of the time constant, i.e., directly proportional to the resistivity. A graph of this form should also allow the time to be extrapolated back to zero time to allow for the initial loss of charge while the meter is being put in place. In practice, it turns out that log Q plotted against t is often not a straight line and no single resistivity can be defined. Only an approximate estimate of the expected charge-decay time can therefore be obtained from surface or bulk resistivity values.

The lack of agreement between charge-decay measurement and resistivity measurement, assuming Ohms' Law is obeyed, is not really surprising. First, Ohms' Law does not apply to conduction in insulating or semiconducting materials or to powders. Second, the movement of artificially applied free charge is likely to differ from that of internal bound charge which is measured in standard resistivity tests.

In many cases, the resistivity is itself a function of the voltage at which it is measured. Higher voltages tend to give a lower value of resistivity because more charge carriers are created at high field levels. The voltage chosen for measurement is somewhat arbitrary but since 5 kV or above is normally required to give a incendiary spark, a voltage around 1 kV is normally chosen. This is low enough to avoid damaging instrumentation if the sample is highly conducting or likely to arc, but high enough to give a reasonable representation of current flow at electrostatic potentials.

3.3.6.4. *Resistivity of Powders*

A powdered material conducts electricity by a complicated path between particles, and the resistivity depends on the surface and bulk resistivity of the solid, the contact area of the particles, and any surface films present on the particles. Resistivity therefore depends on the packing density and the pressure applied to the powder, as well as on particle size and moisture content. It is possible to use resistivity measurements to determine fundamental powder properties such as oxidation and moisture content, but for hazard evaluation a rough classification of resistivity as a guide to charge retention is all that is required. For this purpose resistivity is measured in a cell containing electrodes of known size and spacing. Although resistivity is a function of packing and pressure, for most practical purposes, it is sufficient to simply vibrate the cell gently to achieve uniform packing. A powder resistivity cell is illustrated in Figure 3.19. The electrodes have rounded edges to prevent field intensification and guard rings are fitted to reduce leakage via the cell. A potential is applied between the electrodes and the current I measured. The resistance R is given by $R = V/I$ and the resistivity by $\rho = RA/d$. It has not been found possible to measure powder resistivities above $10^{15}\ \Omega \cdot$ m. A resistivity of this magnitude requires measurement of only 10^{-11} A at an applied potential of 10 kV, and lower currents are difficult to detect without extensive screening. For evaluation of electrostatic hazards it is normally sufficient to divide powders into three groups:

(i) $\rho < 10^{9}\ \Omega \cdot$ m—no charge storage when powders in stationary contact with earth;

(ii) $\rho > 10^{12}\ \Omega \cdot$ m—long-term charge storage for at least several minutes and perhaps at the highest resistivities for several days;

(iii) intermediate range $10^{9} < \rho < 10^{13}$—high velocities lead to sufficient charge production for charge accumulation.

Typical powder resistivities are given in Chapter 5. The dependence of the resistivity of powdered PVC on potential is shown in Figure 3.20. Not all powders show as strong a dependence on voltage as this example, but a sharp increase in resistivity when the applied potential is reduced below 1 kV is not uncommon.

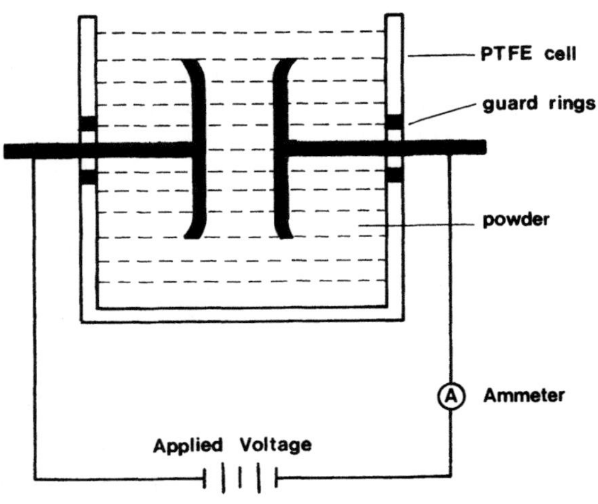

FIGURE 3.19. Resistivity cell.

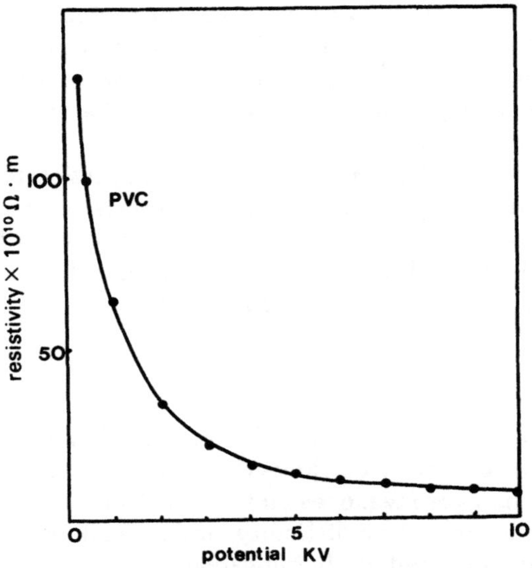

FIGURE 3.20. Resistivity of PVC powder as a function of applied potential across electrodes 5 mm apart.

Figure 3.21 shows the charge on the surface of a layer of PVC. powder as a function of time. The logarithm of the charge is plotted against time and it can be seen that a straight line is not obtained. In this case the initial rapid decay was approximately equivalent to the resistivity measured at low voltages (even though the field across the layer was actually high). The steady charge-decay rate which occurred after the

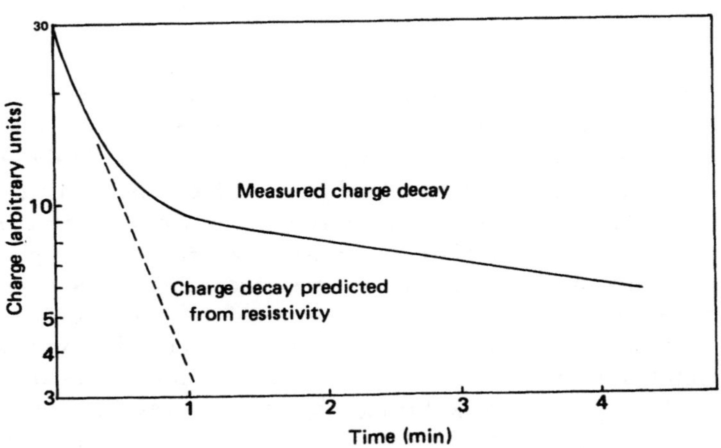

FIGURE 3.21. The charge on the surface of a layer of PVC powder as a function of time.

first minute was equivalent to a time constant of 420 s. A resistivity of approximately $10^{12}\ \Omega \cdot m$ (the maximum measured even at low fields) would be expected to give a charge-decay time of only 10 s. PVC therefore retains charge for longer than would be expected from the resistivity measurement.

Some highly insulating powders show the reverse effect and charge decays faster than predicted. It can be seen that resistivity provides only a rough guide to charge-decay times for a powder sample, sufficient only to classify the powder broadly into the three groups. If charge-decay time is required precisely it must be measured directly.

3.3.6.5. *Resistance Path to Ground*

The resistance of a path to ground is measured by applying a voltage and measuring the current that flows through the meter to the ground. A multimeter or megger measures directly in ohms. The grounding of a metal plate should present no difficulty, and a check with a low applied voltage (a few volts) should easily indicate less than 10 Ω. However, the resistance of grounded path may rise to $10^6\ \Omega$ without causing electrostatic problems. In some cases, for example, where grounding is achieved through a concrete floor, a resistance of this level may need to be measured.

In this case 500 or 1000 V is applied, and the measurement is made with a megger. This instrument is normally used for checking insulation in electrical wiring and will be a common item of test equipment in most industrial plants. The maximum resistivity measurable by a megger is normally $10^8\ \Omega$.

3.3.6.6. *Personnel Resistance Detector*

Personnel have sufficient capacitance to store enough charge to give an incendiary spark. Thus they are generally required to have a resistance to ground of 1 MΩ or less, if they work in an inflammable atmosphere. The circuit shown in Figure 3.22 can be used to avoid applying a high voltage to a man but still obtaining a measurable current. The person stands on the plate and grasps the handle to complete the circuit. The bias of the transistor is altered according to the resistance between the handle and the plate. The output of the transistor therefore changes. The output is amplified and can operate a relay that switches off a green light and switches on a red light if the resistance rises above the set level. The current passing through the person is less than 100 μA and cannot be detected by him. These instruments can be set up so that plant

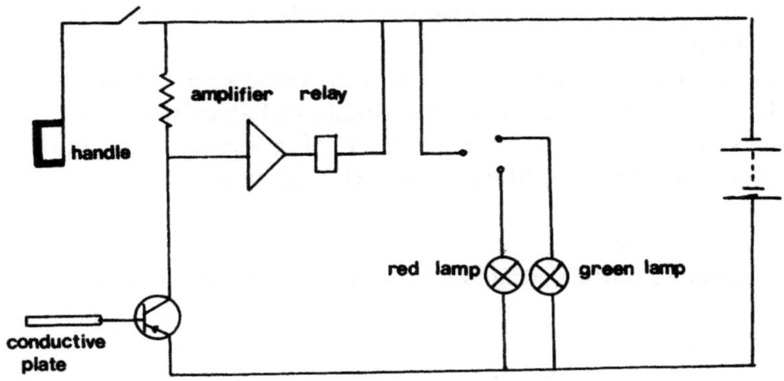

FIGURE 3.22. Schematic circuit for personnel resistivity tester.

operators routinely check the resistance of antistatic footwear which may become dirty and thereby less conducting.

3.3.7. Energy of an Electrostatic Spark

Flammable dusts and vapors are classified as electrostatic hazards by their minimum ignition energy. Therefore, it is necessary to be able to measure spark energies both for the purpose of classifying materials and to evaluate a potential hazard in the workplace.

The electrical energy flowing at any time is the current flowing through the spark gap multiplied by the voltage across it. The total energy of the spark is therefore the current multiplied by the voltage integrated for the duration of the spark. The difficulties of measuring high-voltage transients has been discussed previously and available methods are not very accurate. The current flowing during the spark is easier to measure and can be recorded directly on an oscilloscope. This gives useful information about the shape and duration of the spark and the total charge transferred, even when no voltage measurement is available to give the spark energy. For practical industrial processes, where it is necessary to evaluate spark energy, this technique is far too sophisticated to be practicable. Usually in industry the two ends of the spark are not both accessible. Optical techniques in which the spark energy is related to the light emitted are being developed, but the light intensity is not dependent solely on the spark energy and considerable development is required before this will give a practical instrument. Sparks also emit radiation in the radio-frequency part of the electromagnetic spectrum. So far, no correlation has been found between the detected signal and the spark energy, but radio receivers are very useful in detecting the presence of a spark when interference from machinery is not too high.

Optical and electrical measurements of spark energy may be made in the laboratory to evaluate the explosibility of hazardous materials but in an industrial situation the energy W of a spark must normally be estimated from the capacitance using the relationship

$$W = \tfrac{1}{2}CV^2 \tag{3.20}$$

Since the capacitance voltage and charge are related by the equation

$$C = \frac{Q}{V}$$

the relationship for energy may be written

$$W = \tfrac{1}{2}CV^2 = \tfrac{1}{2}QV = \tfrac{1}{2}Q^2/C \tag{3.21}$$

A knowledge of any two of the three parameters allows the energy to be estimated.

3.3.8. Capacitance

The electrical capacitance of any object defines the potential to which it will rise when it contains a given charge. A capacitance can be measured very accurately using an AC bridge. A simple bridge is shown in Figure 3.23. The variable capacitance and resistor C_2 and R_2 are varied until the unknown capacitance C_1 is balanced and the current through the meter is zero. Sophisticated bridges also make allowances for stray inductances. Most bridges require that neither of the terminals of the capacitor is

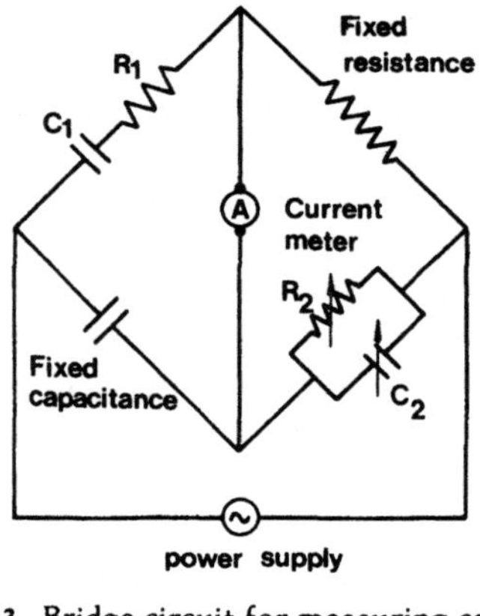

FIGURE 3.23. Bridge circuit for measuring capacitance C_1.

grounded. Electrostatically, it is required most often to know the capacitance between an isolated object and ground, so a bridge is both oversophisticated and unsuitable.

One instrument that can be used in this situation uses the principle that the frequency of an oscillating circuit is proportional to the capacitance in the tuning circuit. Two oscillators are set up at the same frequency. The unknown capacitance is added to one circuit and the frequency difference between the two is counted and displayed on the meter scale. The instrument does not give the very high accuracy possible in a bridge circuit but this is not required for estimates of capacitance for safety purposes. A spark from a capacitance seldom contains all the energy stored and a calculation of $\frac{1}{2}CV^2$ gives the maximum possible energy.

Frequently a capacitance can be estimated or calculated approximately from very simple equations such as given in the two examples below.

(i) Parallel-Plate Capacitance

$$C = \frac{\epsilon\epsilon_0 A}{d}$$

This leads to a very simple relationship between the measured field E and the stored charge since $C = Q/V$ and $E = V/d$; i.e.,

$$\frac{Q}{Ed} = \frac{\epsilon\epsilon_0 A}{d} \quad \text{and} \quad Q = \epsilon\epsilon_0 AE$$

(ii) Capacitance of an Isolated Sphere

$$C = 4\pi\epsilon\epsilon_0 a$$

where a is the radius of the sphere. [The derivations of both these quations can be found in most textbooks on electricity and magnetism, e.g., Bleaney and Bleaney (1957)].

The capacitance of isolated bodies found in industry can be roughly estimated from a sphere of equivalent diameters. Thus a beer can or tool may have a capacitance of 1–5 pF while a man will have 100–300 pF. The capacitance between parallel surfaces is estimated using the parallel-plate approximation.

When two or more capacitances are connected in parallel the capacitances add. When they are put together in series the reciprocals are additive, i.e.,

$$\frac{1}{C} = \frac{1}{C_1} + \frac{1}{C_2} + \frac{1}{C_3}$$

3.4. EXAMPLES OF ELECTROSTATIC MEASUREMENTS IN AN INDUSTRIAL ENVIRONMENT

The relationships between the fundamental quantities in electrostatics are basically very simple. Complications arise only because an allowance must be made for instrumental effects or because the actual parameter required cannot be measured directly. A number of examples are presented below which illustrate particularly difficult situations. In general, it is necessary to combine results from several sources to obtain a self-consistant set of values for charge, voltage, capacitance, and spark energy.

3.4.1. Moving Conveyor Belt

The only instrument that can be used to obtain an estimate of potential or charge on a moving conveyor belt (Figure 3.24) is the electrostatic field meter since this can give a reading without contact. However, this measures field rather than charge. Two separate situations can be defined according to whether the capacitance between the belt and ground is dominated by the presence of a roller or by the presence of the field meter.

Ideally a measurement can be made of the surface potential of the film as it passes over a roller (point A in Figure 3.24) and far from a roller, with the meter placed as close as practical to the belt (point B in the figure).

The capacitance per unit area between the film surface and the grounded roller in position A is given by

$$\frac{C}{A} = \frac{\epsilon\epsilon_0}{d_1}$$

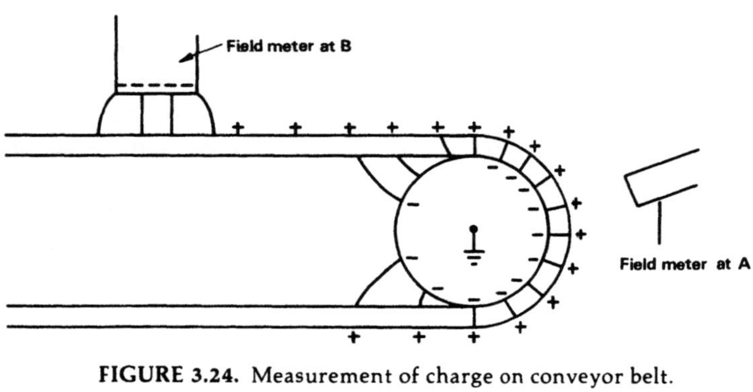

FIGURE 3.24. Measurement of charge on conveyor belt.

where $\epsilon_0 = 8.8 \times 10^{-12}$ F/m, ϵ is the dielectric constant, and d_1 is the film thickness. The charge density Q on the film surface is related to the capacitance per unit area (C/A) by the equation

$$\frac{C}{A} = \frac{Q}{V}$$

Therefore,

$$Q = \frac{\epsilon\epsilon_0 V}{d_1}$$

where V is the surface potential measured with the field meter.

While the belt is directly over the roller the capacitance between the film surface and the meter is negligible compared with capacitance between the film surface and the roller. Thus, there is negligible charge sharing with the meter and the meter reading of surface potential is a true value.

At position B the most important capacitance is that between the meter and the surface:

$$\frac{C}{A} = \frac{\epsilon_0}{d_2}$$

where d_2 is the meter-to-film spacing.

$$Q = \frac{\epsilon_0 V}{d_2}$$

In this case the value of V obtained will be a function of the spacing because larger spacing will allow the stray capacitance between the belt and ground to play a larger part.

Measurements frequently show that the charge per unit area on the belt away from the roller is lower than when the belt is directly over the roller. The voltage away from the roller will actually be higher because the capacitance is so much lower. The decrease in charge density means that a spark discharge occurred as the belt left the roller.

Surface-charge measurement on the belt can be used to evaluate the effectiveness of antistatic dischargers. The charge per unit area multiplied by the velocity gives the total current per unit length of discharger which must be supplied for effective elimination. Conversely, a passive discharger may be used to calibrate the potential measured on the belt in terms of charge density. A passive discharger is an array of sharp grounded points which form a corona discharge when a high potential from a charged surface passes underneath them. If the points are grounded via an ammeter the current drawn by the points can be measured. This is equivalent to the total charge per unit time removed from the belt. If the area of belt passing under the dischargers per unit time is measured then the charge reduction per unit area can be calculated and compared with the reduction in belt potential due to the dischargers which is measured with the field meter.

3.4.2. Energy of a Spark from an Ungrounded Bin of Conducting Powder

An ungrounded bin of conducting powder can be considered to be an isolated charged conductor. The charge on the powder may be measured by catching a sample of powder as it falls from the shute in a Faraday cup such as described in Section 3.3.4. The potential to which the bin rises may be measured with an electrostatic voltmeter or a field meter. The electrical capacitance of a bin containing 100 kg might be ~ 20 pF. An electrostatic voltmeter of 2-pF internal capacitance thus introduces a 10% error, which is acceptable.

The field meter held a few centimeters from the bin has negligible capacitance and the potential on the bin is found by multiplying the measured field by the distance of the meter from the bin. If all the charge entering the bin is stored and none is lost by a spark or leakage then the voltage expected may be calculated.

Let us assume as an example a moderate level of charge on the powder, equivalent to 5×10^{-8} C/kg. The total charge in the 100-kg bin is then 5×10^{-6} C. If the bin has a capacitance to earth of 20 pF then the voltage on the bin would be

$$V = \frac{Q}{C} = 250 \text{ kV}$$

In practice, a spark is likely long before this potential is achieved. The meter will measure a potential that is seen to rise slowly then fall rapidly back as a spark occurs.

The maximum energy W of the spark can be calculated because the capacitance of the bin and the voltage at which the spark occurs are both known. If $C = 20$ pF and the maximum bin voltage is 15 kV the energy is given by

$$\begin{aligned} W &= \tfrac{1}{2}CV^2 \\ &= \tfrac{1}{2}(20 \times 10^{-12} \text{ F})(225 \times 10^{6} \text{ V}^2) \\ &= 2.2 \text{ mJ} \end{aligned}$$

Since all the charge stored in the bin will not be released at once, this spark will not ignite most powders.

If the bin was isolated more effectively so a potential of 25 kV could be built up before a grounded path was found then

$$W = \tfrac{1}{2}CV^2$$
$$= \tfrac{1}{2}(20 \times 10^{-12}\ \mathrm{F})(625 \times 10^{6}\ \mathrm{V}^2)$$
$$= 6.25\ \mathrm{mJ}$$

The energy stored if there was no discharging at all until 250 kV is reached would be

$$W = \tfrac{1}{2}CV^2$$
$$= \tfrac{1}{2}(20 \times 10^{-12}\ \mathrm{F})(625 \times 10^{8}\ \mathrm{V}^2)$$
$$= 625\ \mathrm{mJ}$$

3.4.3. Measurements in a Powder-Storage Silo

Powder-storage silos frequently are accessible only by a manhole or vent cover in the roof. The most useful measurement instrument inside the silo is again a field meter, but no instruments can be inserted during the filling of a large system because of the force of falling powder. It is also inadvisable to insert instrumentation permanently because the field is intensified around a grounded object. A sudden increase in the space potential could lead to a sharp field intensification at the meter, thus causing a spark.

A field meter that is flush with the walls and roof provides no field distortion and will not be damaged, but reads the field due to the integrated effect of all the charges within the bin and does not give a clear picture of the charge distribution at different positions.

If there is an access point well clear of the falling stream, a meter may be lowered but if the field rises above 1×10^5 V/m it should be removed. The field read by a field meter will be due to the combined effect of the potential on the settled heap of powder, the falling stream of powder, and the dust cloud. It is possible that these three regions will contain different charge densities and even different polarities of charge.

A full picture of the electric potential and field distribution in the silo requires field measurements at a number of positions to be combined with estimates of the charge density in different regions and the known boundary conditions of the grounded silo wall. This is achieved most easily by a finite-element computer program which solves the basic electrostatic equation (Laplace's equation) relating field, charge, and potential for every element of the silo's volume. The program can take account of the distortion in measurements produced by the field meter itself when it is not situated in a grounded wall.

Field measurements on the axis of the silo and measurement of the surface potential of the heap of powder are particularly useful. Even without the computer useful estimates can be made from measurements of the electric field in as many positions as the geometry of the system permits, together with measurements of charge density using the field meter surrounded by a grid cage as described in Section 3.3.2.

4

Electrostatic Ignition of Powders

4.1. ELECTROSTATIC CHARGING OF POWDERS

Powder particles nearly always carry a measurable charge because of the contacts made during their production or movement. The charge on a powder particle is governed by three factors: (i) the charge production rate, (ii) the charge leakage rate when the particle is in contact with a ground, and (iii) the electrical breakdown of air initiated by the high field around the charged particle.

The rate of charge production is difficult to predict since it depends critically on the surface properties of the particles and on the materials with which they are in contact as well as on the speed of the frictional contact. The charge leakage rate τ can be predicted theoretically because it is related to the powder resistivity according to the equation

$$\tau = \epsilon\epsilon_0\rho$$

In fact ρ (resistivity) is not usually a constant for a powder and depends on stored charge or applied voltage, but an approximate time constant can be estimated from the resistivity measured at a fixed voltage using the method described in Chapter 3. Typical resistivity values and associated time constants are given in Table 4.1. Even fairly conducting materials such as aluminum have a time constant that may be significant compared with the time taken for a particle to move a distance of one particle radius. The net result of the rapid movement of a conducting particle in a grounded metal pipe is an increase in charge on the powder. As soon as motion ceases the charge decays away from a conducting powder that is in contact with a ground, but if powder is collected on an isolated or insulating surface or is dispersed in the air then charge cannot flow and the powder may retain a high charge.

The maximum charge that a single particle can hold is limited by the electric field created by that charge. This cannot exceed the air breakdown limit because if the field starts to rise above this level the air ionizes

TABLE 4.1. Charge Decay from Powders of Different Resistivities.

Material	Resistivity ($\Omega \cdot m$)	Approximate charge-decay time constant (sec)
Iron	10^4	10^{-7}
Aluminum	10^7	10^{-4}
Grain dust	10^7–10^9	10^{-4}–10^{-2}
Sugar	10^{11}	1
Dried milk	10^{11}–10^{13}	1–100
Nylon	10^{13}	2–100
Polyethylene	$>10^{15}$	$>10^4$

and the free ions created reduce the charge on the particle. The charge limit can be calculated using the relationships discussed in Chapter 3. The field created by a charge q on a particle is given by Gauss's Law, which was discussed in Section 3.2.3:

$$\int E\, dS = \frac{\Sigma q}{\epsilon_0}$$

If the charge per unit volume is Q on a sphere of radius r then the above equation becomes

$$E(4\pi r^2) = Q\,\frac{4\pi r^3}{3\epsilon_0}$$

$$E = \frac{Qr}{3\epsilon_0}$$

Since E must not exceed the breakdown field E_b, the maximum charge density is given by

$$Q_{max} = \frac{3\epsilon_0 E_b}{r}$$

and the total-charge to mass ratio is

$$\frac{Q_{max}}{m} = \frac{3\epsilon_0 E_b}{Dr}$$

where $m = 4/3\pi r^3 D$ and D is the material density. For a 30-μm radius particle this works out to be approximately 2.6×10^{-3} C/kg.

The charge on a particle is proportional to its surface area. The mass on the other hand is proportional to volume. The charge-to-mass ratio is therefore inversely proportional to particle radius; i.e., fine particles have a much higher charge than the same mass of a coarser material.

When the powder particles come together, during transport in a pipe or during storage, the charge must be lower than this maximum because the field at the outside of the volume of powder due to all the particles together rises. This can be seen by applying Gauss's Law to the powder in a pipe.

Assume the powder travels with a density of P kg/m^3. Then applying Gauss's Law

$$\int E\, dS = \frac{\Sigma q}{\epsilon_0}$$

where q is now the total charge in length l of the pipe of radius R and dS is the surface area of this length of pipe. We then have

$$E(2\pi Rl) = \frac{Q(\pi R^2 l)}{\epsilon_0}$$

where Q is the charge/unit volume in the pipe.

$$Q_{max} = \frac{2E_b\epsilon_0}{R}$$

The charge-to-mass ratio is thus given by

$$\frac{Q_{max}}{P} = \frac{2E_b\epsilon_0}{RP}$$

In most industrial situations this limit leads to a lower charge on the powder than would be feasible for an individual particle, so industrial charge levels achieved by friction do not reach the very high levels possible when a dispersed cloud is charged by corona. However, when R or P is small, i.e., the powder is well dispersed or very little is moved, very high values of charge may be achieved.

Table 4.2 shows typical values of the charging levels of organic powders measured in different industrial processes. Since the charge

TABLE 4.2. Typical Charging Levels Found on Organic Powders During Industrial Processing.[a]

Industrial Process	Charge (C)
Sieving	10^{-9}–10^{-11}
Pouring	10^{-7}–10^{-9}
Scroll feed transfer	10^{-6}–10^{-8}
Grinding	10^{-6}–10^{-7}
Micronizing	10^{-4}–10^{-7}

[a] Gibson (1981).

produced in practice is limited by breakdown or leakage and not by the rate at which is produced, the values measured in industrial processes are a stronger function of the process than of the material.

For any given process it may be necessary to estimate which product is likely to give the worst electrostatic conditions. In general terms, resistive powders are obviously the most likely to store charge but if the powder is to be dispersed so its charge cannot leak it may be valuable to make an actual estimate of powder chargeability.

A number of attempts have been made to measure this parameter in the laboratory. These experiments are designed as a comparison between powders. So far, no tests have successfully predicted the charge levels that will be obtained in a particular industrial process. In the simplest laboratory test, powder is passed through a straight tube. Some workers have used pipes of 18-mm diameter with a powder density of ~0.05 kg/m^3 in air (Boschung and Glor, 1980) while others have used a smaller bore pipe of 5–10 mm diameter and a powder density of 2 kg/m^3 (Morris, 1978). In the former case the powder density is sufficiently low for the limitation on charge to be set by chargeability and leakage. In the latter case surface discharges at the pipe wall may occur.

Industrial charge densities in dilute-phase pneumatic transport are generally between 0.1 and 1 kg/m^3 and the pipe is large so that discharging is more probable. High-density tests are therefore a better representation of industrial practice but results tend to be less repeatable and more dependent on apparatus geometry. The high-density measurements reflect both chargeability and the discharge characteristics of the powder but not necessarily in the same way as in an industrial process. The lower-density measurements measure chargeability more accurately but do not take account of the discharge processes. Boschung and Glor in low-density measurements showed that, although the charge produced was affected by the material of the tube through which the powder was pumped, the variation between different powders was much higher than the variation with tube material. The charge produced was primarily a function of the nature of the powder itself. A broad classification into low, medium, and high charging tendency was feasible in spite of variations between tests with different tubes and velocities.

The second class of experiments attempt to put as high a charge as possible on the powder using a rotating fan or disk. Powder is thrown outwards by the centrifugal force and scraped along the surface of the fan. A high charge tends to develop and can reach the theoretical maximum for a single particle. The straight tube in the low-density mode usually achieves ~ 10% of the maximum possible charge. An interesting feature of the fan type of charging device is that bipolar charging has

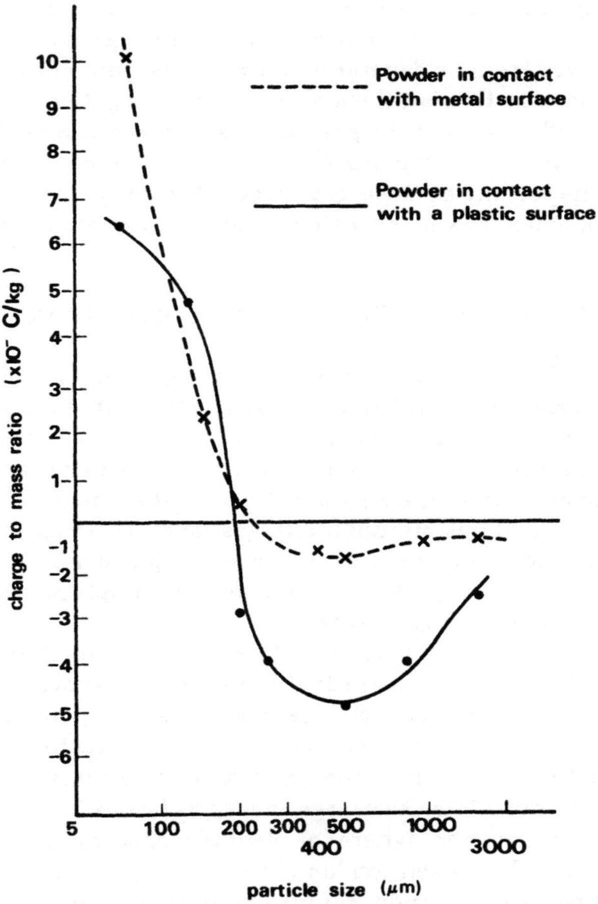

FIGURE 4.1. Particle charge as a function of particle size (fan charging device).

been observed (Cross, 1979). That is to say when particles of the same material but different size ranges are passed through the apparatus there is a dependence of charge magnitude and polarity on particles size (Figure 4.1). In the case shown, the small particles were positive and the larger particles were negative. The same effect was not observed with the same powders tested in straight pipes. It is possible that bipolar charging is caused by charge transfer in interparticulate contacts of different-sized particles of powder, rather than by contact with the walls of the pipe through which it is flowing. Hence bipolar charging will be affected by turbulence and the way the powder flows.

Bipolar charging has also been observed in an industrial silo. Again, the fine particles were positive and the coarse negative. The phenomena was observed from field meter measurements during pulse filling. While the product was flowing the meter showed a negative electric field, but when the filling ceased the polarity changed and slowly decayed as the charged cloud settled. Bipolar charging is by no means always observed but it is important because gravitational settling can lead to a charge separation when bipolar charging occurs as a function of size.

4.2. ELECTROSTATIC CHARGE ACCUMULATION

If an individual dust particle less than 70 μm in diameter is carrying its maximum charge, it is likely to store less than 10^{-15} C of charge. 10^{-7} C is necessary for an incendiary spark (see Section 4.4.2). Therefore, charge must accumulate in some way before there will be an electrostatic hazard. In bulk-powder operations the charge created lies between 10^{-8} and 10^{-5} C/kg. Thus although charges must accumulate from a large number of individual particles the mass of powder needed to provide enough charge for a spark is not high, even in simple pouring operations that do not create a high charge.

The most common means by which charge accumulates is by storage on isolated conductors. Even if the powder transported is reasonably conducting it will transfer charge to an isolated conductor as it moves across it. This transfer by contact between conductors is often overlooked and occurs because the poor contact between a particle and a surface does not allow the charge to leak as fast as it is created.

In many operations where the plant is normally grounded an isolated conductor may have been accidentally created by use of a plastic material. For example, antivibration mountings must be bonded across if they isolate sieves or other moving parts. Flexible hoses sometimes isolate conductors and even the powder itself is an isolated conductor when a charged conducting material is poured into a polyethylene bag. Once charge has collected in this way the whole capacitance of the conductor can discharge in a single spark. The energy of the spark is given by $\frac{1}{2} CV^2$ where C is the electrical capacitance to ground the V the voltage to which it rises. Table 4.3 shows typical capacitances of objects that may become isolated and the charge and voltage that must be stored to be equivalent to 5 mJ of energy, which will ignite some of the more static-sensitive dusts.

When a plastic surface becomes charged it does not normally discharge in a single spark. The slight crackling heard when a sheet of

TABLE 4.3. Hazardous Charge and Voltage Accumulation

Object	Capacitance (pF)	Charge for 5 mJ ($\times 10^{-7}$ C)	Voltage for 5 mJ (kV)
Small scoops, beer can, tools	5	2	45
Buckets, small drums	~20	4.5	22
Large container (50–100 gal)	~100	10	10
Man	~200	17	7
General plant	1000	30	3

plastic is rubbed rapidly is caused by microscopic discharges from very small areas of plastic. The maximum charge that can be applied to an isolated insulating surface without discharge occuring is $\sim 3 \times 10^{-5}$ C/m^2. (This can be proved from Gauss's Law by the method used in the calculations in Section 4.1.) At this charge level the surface is insulating and charge does not spread across it. If the charge on the surface is artificially increased, then the surface can become much more conducting and charge can spread to the spark center from a large area, giving a Lichterberg discharge of high energy. This type of discharge is sufficiently energetic to ignite a dust (Heidelberg, 1967).

The limit to the surface charge is fixed by the charge-created electric field above the surface. One way in which an artifically high surface charge can be produced is to back a layer of plastic with a grounded metal sheet. This effectively forms a capacitor. The field is then built up across the plastic (which will have a relatively high breakdown strength) and the field in the air above the plastic is reduced. Experiment has shown that the plastic must be less than 8 mm thick if the capacitance is to be large enough to give a Lichtenberg discharge (Heidelberg, 1960).

It is extremely difficult to estimate the energy of a spark from a plastic surface. The charge transferred in the spark comes from a rather ill-defined area and the voltage changes give little guide since charge will spread across the surface to discharged regions. The charge transferred during a spark from an insulator surface may be measured directly on an oscilloscope or transient recorder by measuring the voltage developed across a low-value resistor between a sphere and ground, but the voltage change is not measurable. The energy must be estimated by comparing the ignitability with a standard capacitance spark using different gas/air mixtures.

Sparks from plastic surfaces without metal backing will ignite hydrogen and similar gases and can, under some circumstances, produce sufficient energy to ignite gases of the propane class to which most flammable solvents belong. There is a strong argument to suggest that all highly insulating plastics should be eliminated in the vicinity of flammable vapors and hybrid mixtures of vapors and dusts. This may be impractical and where plastics must be used they should not be in a position where they will be dry and involved in frictional contacts.

As laboratory tests proceed, the energy produced from a plastic surface increases and the ignition energy of dusts is being pushed down. Sulfur and aluminum, for example, have been ignited with sparks of less than 1 mJ produced by a specially designed spark source described by Eckhoff (1975, 1976), and sparks of up to 1 mJ have been produced from plastic surfaces. However, as yet there is no evidence to suggest that sparks from isolated plastics have suitable characteristics to ignite dust clouds. In a situation where a plastic can both receive a very high charge and has grounded backing, high energies of at least 100 mJ may be able to be produced and powders may be ignited (Hughes *et al.*, 1975).

Artificially high charges can also be produced on plastics if highly charged sheets are stacked together. This is found when a charged web is wound onto a roller, when charged sheets are stacked, or when highly charged insulating powder particles are piled together during storage. It has been found by Maurer (1979) that large energetic sparks can occur when piles of highly charged powder build up during storage as the central high potential region slides towards the grounded wall.

The ways in which charge can accumulate on isolated conductors or plastics from individual charged particles are well known. Accumulation in the powder itself is less well understood. When a highly charged powder is dispersed in a cloud it produces an electric field. If the field exceeds the air breakdown limit it is theoretically possible for the air to ionize, producing a conducting path through which the cloud can discharge in a lightning-type spark. The air breakdown limit is approximately 3×10^6 V/m. Fields significantly below this will not cause air breakdown. A probe or meter inserted into a charged cloud will cause an intensification of the field near the probe, thus fields below the breakdown limit can rise to dangerous levels on insertion of a grounded projection.

Work in propane gas on the ignitability of sparks created by highly charged mists has shown that the minimum space potential at the point of insertion of the probe must be 70 kV if there is to be a spark that will ignite the propane gas. It may be assumed that *at least* this space potential would be required to ignite a dust cloud.

Charged powder clouds have never in fact been proved to have initiated an explosion. Experiments by Boschung *et al.* in Switzerland (1977) artificially produced a very highly charged dust cloud in a 3-m diameter chamber and failed to obtain a spontaneous explosion in 50 attempts. They concluded that sufficiently high fields could not be achieved in containers as small as this.

In these experiments only unipolar charge was considered and the effect of ions created by powder discharges during settling was also ignored. With an insulating powder, very high fields build up in the powder as it settles. These are reduced by microscopic air discharges creating ions that are likely to increase the tendency of the air to breakdown. Projections within a highly charged powder cloud are also likely to produce corona ions forming a conducting path along which major discharge may occur. This is the mechanism used in lightning conductors to encourage the lightning to strike the conductor rather than the building to be protected.

The effect of the ions could be either favorable or unfavorable. They could either produce a conducting path along which a spark might form or, because they move rapidly in an electric field, they could establish an equilibrium within the chamber to prevent the build up of very high fields. Experiments have not yet established which process is more important.

Charge can also accumulate on powder if a conducting powder settles in such a way as to create an isolated conductor. This may be possible on insulating pipes or filter cloths, but it is unlikely in either case that sufficient energy or charge will be stored in each isolated patch of powder to ignite most dusts. Powder poured into insulating containers can accumulate charge and may form a source of ignition of considerable energy.

4.3. ELECTROSTATIC SAFETY CRITERIA

4.3.1. Energy

The parameter by which the explosibility of a powder is defined is the minimum ignition energy. This quantity has been measured for a large number of different dusts and results are published and readily available. Unfortunately the ignitability is not solely a function of the spark energy. The duration of the spark is certainly important, with longer sparks igniting a dust more easily than shorter-duration sparks. It is possible that the spark shape is also relevant. The result is that different workers using different electrical circuits to produce their sparks achieve widely differing values of ignition energy. Thus, Eckhoff (1975, 1976) using a circuit tuned for maximum ignitability, achieves a minimum ignition energy of 7 mJ for lycopodium powder, whereas the

U.S. Bureau of Mines quotes 40 mJ. A simple capacitive discharge through a 15 kΩ resistor gives a value of 30 mJ. The different circuits and dust-dispersion systems used for this test are described in Chapter 5.

The minimum ignition energy is a strong function of particle size. Lycopodium thus provides a good standard because it has a well-defined particle size and shape. The size distribution of the powders tested are seldom quoted and this, together with uncertainties due to the differences in laboratory testing techniques, means that tables of minimum ignition energies can only be treated as a guide of ignitability and not as strict safety limits.

Eckhoff (1976) has now managed to ignite some more-sensitive dusts, such as 5-μm aluminum and sulfur dust, at energies of 1 mJ or less (aluminum 1 mJ, sulfur 0.3 mJ). However, it is arguable that the circuit which produces these very low levels would never occur naturally. (The United States Bureau of Mines quotes 15 mJ for both these materials.) The Bureau of Mines circuit certainly overestimates the minimum energy because the loss in the transformer in the circuit is ignored. Typically transformers are ~20–30% efficient so the U.S. Bureau of Mines figure might be expected to be a factor of 3–5 too high.

The energy of a spark in industry can be estimated from the capacitance of the object storing the charge using the equation

$$W = \tfrac{1}{2}CV^2 = \tfrac{1}{2}QV = \tfrac{1}{2}C^2/Q$$

Laboratory minimum-ignition-energy measurements can be used to estimate the type of spark that would be required to ignite the dust.

A man charged to 15 kV will give a spark of ~17 mJ. Powders whose ignition energy (measured with a capacitor spark) is greater than 25 mJ are unlikely to be ignited by small static sparks of this type, but major static sparks on isolated conductors or metal-backed plastics which can rise beyond 20 kV may give energies up to a few hundred millijoules. Welding torches and flames can provide energies of many joules and will ignite most flammable powders.

Hybrid mixtures of powders and flammable vapors must be considered to have the ignition energy of the vapor, so sparks of less than 1 mJ must be avoided. Pouring powders from polyethylene bags into flammable solvents has been shown to ignite the mixture on a number of occasions.

4.3.2. Charge

The total charge transferred in a spark can be measured in the laboratory by an oscilloscope as shown in Figure 4.2. It has been shown that a spark which will ignite a flammable vapor (energy 0.2 mJ) contains

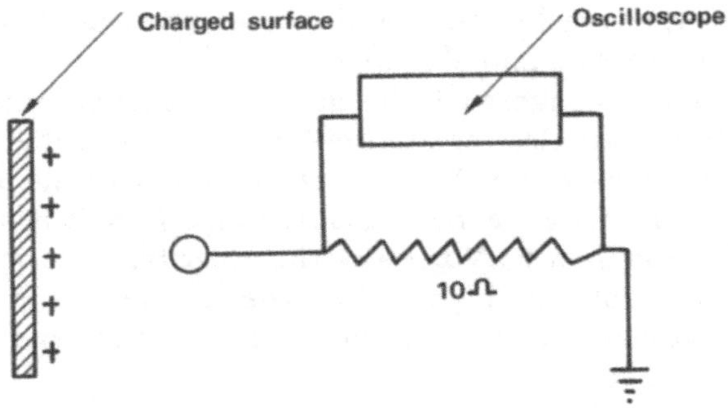

FIGURE 4.2. Measurement of charge transferred in a spark.

at least $5 = 10^{-8}$ coul of charge. This figure can also be taken as a safety limit for dusts, allowing a good safety margin.

The total charge produced by an industrial process can often be estimated from field meter measurements or found directly using a Faraday cup. It nearly always exceeds 10^{-7} C if any of the materials involved is not a perfect conductor. It is seldom practical to prevent charge formation or to keep total charge production effectively below this level. The major effort must be to prevent excessive accumulation of the charge before it leaks to a ground. For conductors that is achieved by effective grounding.

On a highly insulating material, surface charges are normally limited to $3 = 10^{-5}$ C/m^2—a charge level hazardous only to vapors. However, if measurements show the accumulated charge has a density an order of magnitude higher than this, then Lichtenberg discharges may occur. In this case, either the insulator should be replaced by a conductor or some form of static elimination should be applied.

4.3.3. Voltage

It has been shown in experiments with gases that a spark can only ignite the gas if the distance between the metal surfaces, between which the spark occurs, exceeds a minimum distance known as the *quenching distance*. When the metal surfaces are closer than this, most of the heat generated in the spark is absorbed by the metal surfaces. The quenching distance is not a well-defined parameter for powders but electrode spacing must certainly exceed 2 mm (the quenching distance for gases) and normally exceeds 4 mm. A 2-mm gap between spheres of 2-cm radius in dry unionized air requires 8 kV for breakdown. A good safety margin is normally allowed on this figure in order to take account of early breakdown due to ionization at sharp edges, moisture, or ultraviolet light. However, it may be considered that a static potential of less than 4–5 kV will not give an ignitable spark.

There are two dangers which arise from using potential measurements to define a safe electrostatic limit. First, the potential is a function of capacitance; therefore if the capacitance decreases the potential will rise. For example, the potential measured on a film as it passes over a roller may be less than 5 kV because the capacitance between the film surface and the roller is high. When the film leaves the roller, the capacitance decreases drastically, the potential rises, and a spark will occur back to the roller.

Second, a problem occurs if there is a current source constantly supplying the potential. An extreme example is the spark from a 12-V car battery which contains a high energy at low potential because there is a large current source. The peak current in an electrostatic spark is at least 1 mA; therefore the 4-kV potential limit may be used only if no current supply exceeding this level is available.

High potentials are not invariably hazardous. For example, it is not unusual to measure 20 kV or more on a plastic surface that has a low capacitance because it is relatively far from a grounded surface. The small area that discharges in any spark contains little energy. Thus, although the safe limit of voltage for conductors is exceeded, the available charge is below the level required and no incendiary spark is produced.

No one hazardous potential can be defined for a plastic and it is necessary to estimate the surface charge density to see whether high-energy Lichtenberg discharges could occur.

Liquids present an intermediate case. Charge is distributed throughout the liquid and requires a finite time (on the scale of the duration of a spark) for the charge to spread through the liquid to the spark point. Measurements have shown that a potential of 25 kV can be tolerated on a fuel surface before a spark can be obtained from the surface which would ignite the vapor. Again no measurements have been performed specifically with powders also present, but the 25-kV limit may be assumed also to apply to hybrid mixtures of powders and vapors.

4.3.4. Resistance

The maximum resistance of the grounded path that may be allowed if static accumulation is to be avoided is internationally agreed to be 1 MΩ. This limit will provide a good leakage path in all situations no matter how high the rate of static-charge production. It provides a large safety margin in most common situations.

Major plant items must in any case be grounded directly because of regulations concerning electrically operated equipment, but passive items set in concrete may have resistance to ground of the order of several MΩ. Direct grounding of a metal plant will give a resistance path to ground of the order of 1 Ω. Some regulations state a maximum allowable resistance of 10 Ω because resistances above this level on direct connections imply a faulty connection that could become an open circuit. However, it is not actually necessary to have a resistance as low as this to prevent static accumulation. In fact, the maximum resistance that can be tolerated depends on the rate at which charge is supplied. An incendiary spark from an electrostatic source requires a potential difference exceeding 4 kV. Allowing an order-of-magnitude safety margin on this figure the British Standard Institute code of practice recommends that a maximum of 300 V can be tolerated. The allowable resistance R can therefore be calculated from the rate of charge production or current I by Ohms' Law.

$$V = IR$$

For a voltage limit of 300 V and a current which might lie between 10^{-4} and 10^{-11} A, the acceptable resistance thus lies between 10^{6} and 10^{13} Ω. 10^{6} Ω will be safe in all electrostatic situations but higher resistances to ground may be tolerated if the rate of arrival of charge is known to be low. It is unusual for processes to produce charge more rapidly than 10^{-6} A. Therefore, a resistance of 10^{8} Ω is often adequate to prevent charge accumulation.

4.3.5. Electric Field

The electric field required for a discharge is a function of the gap across which the spark is to occur. Thus at 1 mm a field of 4.7×10^{6} V/m is needed and at 10 cm, 2.4×10^{6} V/m produces a spark. These figures apply to electric fields between two spherical electrodes. A general rule-of-thumb figure is usually taken to be approximately 3×10^{6} V/m. Electric fields below this value should not produce sparks. The introduction of a meter into a powder cloud causes a distortion of the electric field so that the field seen by the meter is increased above the true level. This can produce a hazard unless a careful watch is kept to ensure that fields are well below the danger level. The field measured will be the field between the charged objects (the powder cloud or settled layer) and the meter. Introduction of another grounded object closer to the area of charge will produce an increased field in that area. Field measurements are therefore seldom used directly as safety criteria.

4.4. ELIMINATION OF STATIC

4.4.1. Grounding

The first line of defense against the accumulation of static charge is to arrange for materials to have sufficient volume or surface conductivity for the charge to leak to ground. The time taken for charge to leak is given by RC, where R is the resistance of the grounded path and C the capacitance. This may also be written $\epsilon\epsilon_0\rho$, where ρ is the resistivity of the material and ϵ the dielectric constant.

The basic requirement laid down in international regulations is for a leakage path to ground to be less than 1 MΩ. This may always be achieved with a metal plant, but powders themselves and nonmetallic items may have significant resistivity and store charge. Powders in particular may store charge because the contact between particles is small. Powders are often many of orders of magnitude more resistive than the bulk material from which they are made.

Many products have a resistivity that is a strong function of humidity and operation at a relative humidity of 60–70% can reduce the surface resistivity by four or five orders of magnitude. The humidity required is not excessive and small quantities of stream have even been introduced outside the bag of fluid bed driers to reduce electrostatic problems without adversely affecting drying efficiency. The technique cannot be used with powder that degrade with moisture absorption, and this level of humidity may in any case be too high for the comfort of operators in some situations.

A relative humidity above 65% causes a thin layer of water to condense on surfaces; this is therefore an effective way of introducing conductivity to highly insulating surfaces to prevent charge accumulation. Water can be introduced directly, as in the damp bag which is trailed on continuous sheets of plastic during some printing processes, or indirectly by increasing the ambient humidity.

In some situations charging can be reduced in beds of insulating powder or on insulating surfaces by the use of a mesh of grounding wires. Plastic pipes are also frequently wrapped with grounded wires, fluid bed drier bags may have metal woven threads, and arrays of grounded rods are inserted into storage bins containing highly insulating powders and into fluid bed driers. There is no doubt that the use of closely spaced grounds reduces static. However, it is also possible for an additional hazard to be introduced if they become isolated or if the capacitance of a plastic surface is increased by the presence of the ground.

Grounded rods inserted into a highly charged powder cloud divide up the cloud so that the size of any region that may discharge is reduced.

They also serve to force some of the individual powder particles to discharge directly so that the overall level is reduced. Grounded rods used in this way are not the same as static eliminators described in the next section since they do not produce ions themselves. In general, flexible hoses wrapped with grounded wires are not hazardous although the capacitance of the surface is increased because the flexible PVC from which they are made is usually not insulating enough to be capable of holding the very high charge needed for a spark from an insulator surface. However, the wire is frequently not connected properly to ground and then represents an ungrounded conductor for static storage.

Ground wires woven into fluid bed drier bags and dust filter bags also frequently break during usage, forming small isolated conductors. There have been a number of fires in fluid bed driers attributed to this source, particularly when the wires are of substantial size and solvents are dried. Fibrous bags in which there is a continuous mesh of conducting fibers or bags made of epitropic fibers in which each fiber contains a conducting carbon portion are more satisfactory.

The use of grounding, providing it remains continuously connected to ground, is never in itself dangerous. However, the ground must not be moved into place after the charge has built up and a grounded sheet must not be placed behind an insulating surface that is receiving a very high charge.

4.4.2. Personnel

Operators pouring or sweeping powders, walking across an insulating floor, or removing an outer garment may very easily become charged. The highest potential to which a human body can rise in a dry environment is approximately 50 kV, and this gives a maximum possible spark energy of ~250 mJ. In fact, the whole energy is never dissipated in a single spark and it is unusual for a man to charge higher than 15–20 kV. But sparks from human beings can ignite dust clouds with moderate ignition energies around 25 mJ and will very easily ignite vapors or hybrid mixtures of dusts and vapors which form when powders are poured into solvents.

There are a number of precautions that can be taken. First personnel operating in an area where there could be a flammable atmosphere of less than 100-mJ ignition energy should wear footwear that prevents static buildup. Conducting footwear is normally grounded by direct metal connections but it is more usual to use antistatic footwear which also gives protection against electric shock from mains voltages up to 250 V. Antistatic footwear has a maximum resistance to ground of $5 \times 10^7 \Omega$ (BS 2506). Since a human being has a capacitance ~200 pF the maximum charge-decay time is then 10 s. Antistatic footwear may show an increase in resistivity with use and should be tested periodically. Leather-soled shoes often provide sufficient conductivity to avoid static buildup in any case, and it is also possible to wear pads that clip under a normal shoe and around the leg to make a grounded contact.

If there is a grounded conducting path through footwear, the wearing of nylon overalls is not a hazard in a dust environment. The sparks from the material itself are not sufficient to ignite a dust and removal will not charge the man if he remains grounded. However, the charge on clothing could conceivably be transferred to another isolated conductor so clothing should always be removed in a safe area. An alternative to antistatic footwear in an area where static is merely a nuisance is to design the system so that operators frequently touch a grounded metal or wood. Often a minor nuisance can be eliminated by improved grounding, for example, by removing rubber feet on stools or attaching grounding chains.

4.4.3. Antistatic Materials

Antistatic materials may be produced in a number of ways. Antistatic plastics are normally made by the addition of carbon fibers or carbon powder to the material. As the percentage of carbon present rises, the conductivity of the material initially increases slowly but at a certain percentage (which depends on the carbon-particle size) the conductivity rises rapidly. Antistatic plastics normally have a resistivity between 10^9 and $10^4\ \Omega \cdot$ m and thus are sufficiently conducting to leak static charge.

There is a danger in the production of these materials that the carbon may not be dispersed evenly within the plastic. This can create conducting islands surrounded by insulating zones. Fibrous additives may align so that the material is electrically conducting in one direction but not the other. It is important to check the conductivity of antistatic materials in all three directions over a variety of different areas.

Conducting plastics based on most of the common polymers may be obtained. They are rather expensive and the higher the percentage of additive (and hence the more conducting) the more it is necessary to sacrifice some of the useful characteristics of the plastic—for example, flexibility.

Textiles are frequently made conducting by the use of antistatic additives that are surface-active agents with a strong affinity for moisture. The most commonly used antistatics belong to one of the four groups: (i) amine derivatives, (ii) quarternary ammonia salts, (iii) phosphate esters, and (iv) ethylene oxide condensates. In all of these, water is absorbed and so the surface conductivity is considerably increased.

Unfortunately, by their very nature, these additives are also soluble in water and hence tend to wash out. Any purely surface additive also tends to wear away with age. The effectiveness of this type of additive is affected by the relative humidity, but conductive paths remain down to reasonably low levels of relative humidity depending on the product. Recently, fibers have been developed with a conducting carbon sheath (ICI epitropic fibers). These are particularly useful for bag filters since the additive neither washes out nor breaks with usage.

4.4.4. Electrostatic Dischargers

4.4.4.1. Passive Dischargers

Electrostatic dischargers reduce static electricity by creating ions which increase the conductivity of the air and neutralize the problem charges.

The simplest type of discharger is a passive device that uses the electric field created by the charge itself to produce a corona discharge at a grounded point. A typical passive discharger starts to operate when approximately 1 cm from a 5-kV voltage or when the field at the points from a charged cloud is equivalent to this. A passive discharger cannot reduce the charge to zero since it ceases to operate at moderate charge levels when the field becomes too low to initiate a corona discharge.

Passive dischargers are most frequently used on insulating conveyor-belt systems or moving webs in paper and film making. The use of passive dischargers on belt conveyors of powder is of limited effectiveness because the charge is reapplied when the belt passes over the next roller.

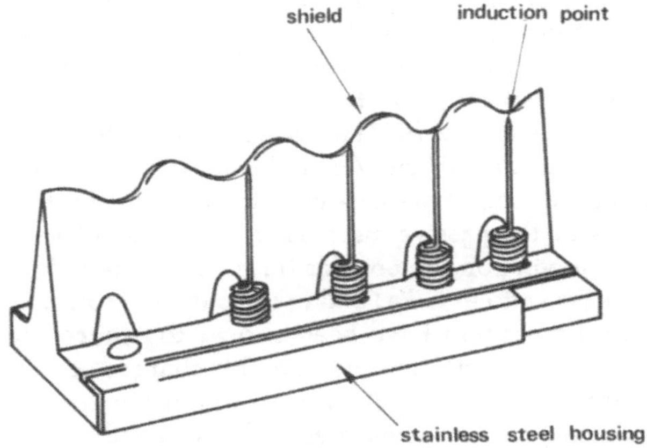

FIGURE 4.3. Passive discharger.

However, they can be useful if there is one particular point where the hazard is high, such as before entry to a solvent.

There are a number of different designs of passive discharger for belt systems ranging from single points separated by approximately 1 cm (Figure 4.3) to a tinsel design with many brushlike metal strands. In a corona discharge from an array of points, the points act independently only when the spacing exceeds 1 cm. Closer spacing of the points does not give a more efficient discharger. A grounded fine wire makes an extremely simple passive discharger effective in reducing charge. The discharge from a wire is seldom uniform and there are usually a number of high-current hot spots irregularly spaced along the wire. Discharging is therefore less uniform than with an array of points.

The threshold field E_c that must be exceeded at the surface of the corona wire before a discharge is initiated is given approximately by

$$E_c = \delta A + \frac{B_x}{(r,\ \delta)^{1/2}}$$

where r_1 is the radius of the wire, δ is the relative gas density with respect to air at 1 atm and 25°C, A = 32.2 (10^5 V/m), and B = 8.46 (10^4 V/m)$^{1/2}$. Values of the threshold field for different wire diameters are given in Table 4.4. The field at the wire is given by

$$\frac{r_2}{r_1} E_s$$

where r_2 is the spacing between the wire and the unwanted charge and E is the field at the charged surface. Although the corona onset field is higher as the diameter of the wire decreases, this is more than compensated for by the field intensification of the smaller radius. The wire is therefore chosen to be as fine as in consistent with mechanical stability.

Passive dischargers may be used to discharge powder entering a storage bin, providing the field shown by a field meter held at the position of the points exceeds 5 kV/cm. At this level of field a current of a few tens of microamperes might be expected from each point. Therefore the effectiveness of the passive discharger can be calculated by comparing the current flowing on the powder (mass flow rate in kg/s multipled by charge-to-mass ratio in C/kg) and in ion current from the dischargers.

Passive dischargers have been shown to reduce high charges by about 50% but are of little use on less highly charged powders because the potential produced is not high enough to initiate discharge. There are also mechanical problems associated with the introduction of protrusions into a powder flow.

TABLE 4.4. Threshold Field for Different Wire Diameters (Passive Discharger).

Wire diameter (cm)	Threshold field (V/m × 10^6)
1.27	4.3
0.88	4.54
0.54	4.94
0.33	5.46
0.92	7.57
0.31	10.7
0.012	15.3
0.0025	29.8

4.4.4.2. Active Dischargers

Active dischargers are of two types both of which actively ionize the air. In one case the air is ionized by α particles from a radioactive source and in the other by a corona discharge created at a point by a high alternating voltage.

Radioactive dischargers most often use a polonium 210 source. The radioactive material is deposited below the surface of a silver foil. The α particles have a range of 3.85 cm in air and the source has a half-life of 138.4 days. It is not the α particles themselves that neutralize the charge but the plasma of ions of both polarities created when the α particles passing through the air collide with air molecules. A typical radioactive bar has an activity of 80 mCu giving a neutralization current of approximately 8 μA/m. A radioactive device creates a diffuse plasma and tends to give more complete neutralization than other active neutralizers, but it cannot cope with high rates of discharge. For high charge levels a radioactive discharger is often used in conjunction with passive devices. Alternatively AC neutralizers are used. A major advantage of a radioactive neutralizer is that it can be easily certified intrinsically safe and can therefore be used in a hazardous environment. However, it is relatively expensive and cannot be used in dirty environments which would entail mechanical cleaning.

In an AC active discharger a potential exceeding 5 kV is applied to an array of sharp points. The points do not need to be extremely fine providing the potential divided by the radius of curvature of the points exceeds the threshold field. The ion current produced by the discharger depends on the potential applied and the proximity of the nearest ground. High currents are produced if the voltage is high and if the mean electric field between the points and ground is high. Alternating current dischargers are normally designed with a wire or array of points situated within a grounded shield (Figure 4.4).

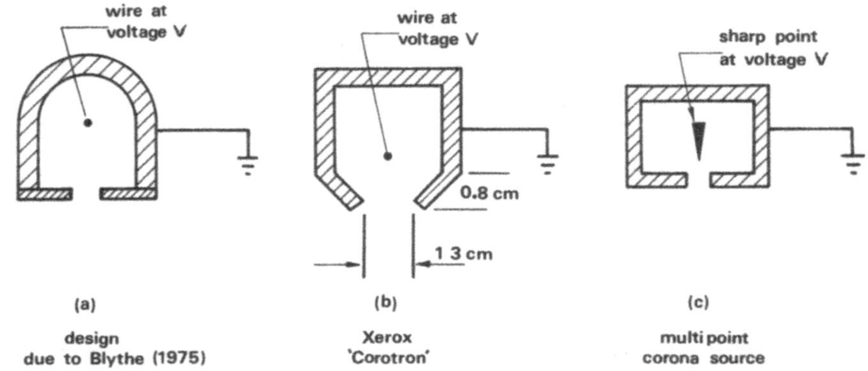

FIGURE 4.4. Alternative geometries for AC high-voltage dischargers.

The corona discharge takes place primarily between the points and ground but ions will be deflected out of the shield to neutralize charged zones. A single point can produce currents up to a few hundred microamperes. Under typical operating conditions one might expect a maximum of ~ 1 mA per meter length of a linear AC discharger.

Although a shield that is close to the wire increases the wire current, it also means that a higher field is required to deflect ions out of the discharger. Commercial devices vary considerably in dimension with shields between 0.5 and 5 cm from the wire. There is an optimum spacing between an AC discharger and the charge to be neutralized. If they are too close the discharger tends to deposit zones of positive and negative charge instead of neutralizing uniformly. If it is far away there is some mutual annihilation of the ion clouds produced on the positive and negative halves of the cycle and efficiency is reduced.

The frequency of the corona source depends on the speed with

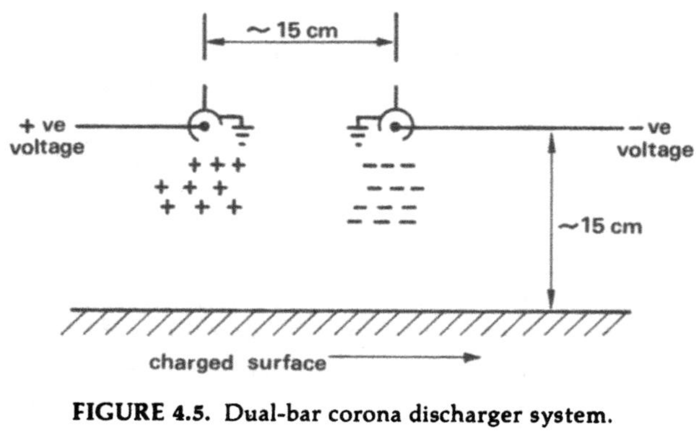

FIGURE 4.5. Dual-bar corona discharger system.

which the charge to be neutralized is moving. At 50 Hz the maximum speed of a charge that can be removed is 2 m/s.

In an AC discharger positive and negative clouds are supplied at different times as the high voltage oscillates. It is also possible to design a dual-bar corona system. In this device, illustrated in Figure 4.5. the positive- and negative-charged clouds are produced simultaneously from two arrays of corona points; one at a positive and the other at a negative potential. The corona points create a weak plasma in front of the charged work-piece from which the appropriate sign of charge for neutralization is extracted. The device operates effectively with much bigger (and less critical) spacing between the problem charge and the naturalizer, but it is more expensive and takes up more room than a conventional device.

Active dischargers are used industrially on the small-scale packing of powders to prevent charged dust from sticking to the outside of tins and containers but are not common on larger-scale powder systems.

Trials have shown that charge can be reduced by an order of magnitude on large-scale transport of powder with powder traveling at 20 m/s. This is typical of most AC discharger applications. Alternating current dischargers are not used commercially on large-scale powder systems. One problem is the necessity of ensuring that the device cannot produce a spark in a fault condition. To do this, it is necessary to have a large resistance between the source of power and the high-voltage point. Then even if a short circuit occurs at the points, the maximum current that can be drawn is set by the applied voltage divided by the limiting resistor. This resistor must never be short-circuited. In addition to limiting the current, it is also necessary that the capacitance between the discharger part after the limiting resistor and ground must not be high enough to allow sufficient charge to be stored to give a spark. This sets a limit on the size of any structure on which a point is mounted. It effectively means that for a large system each point must be separately mounted with a separate limiting resistor. The same effect can be achieved by a low decoupling capacitor instead of a resistor.

An alternative system for producing clouds of positive and negative ions from an AC source is shown in Figure 4.6. The high-voltage electrodes are completely encapsulated in insulating material. Grounded points in close proximity to the encapsulated electrodes give high corona currents when an alternating potential is applied to the electrodes. No spark can be produced from the electrodes because they are surrounded by plastic and their potential is well below the breakdown potential of the plastic. The points are themselves at ground potential and so give no spark on approach of an external ground.

There have been few detailed tests on the relative effectiveness of

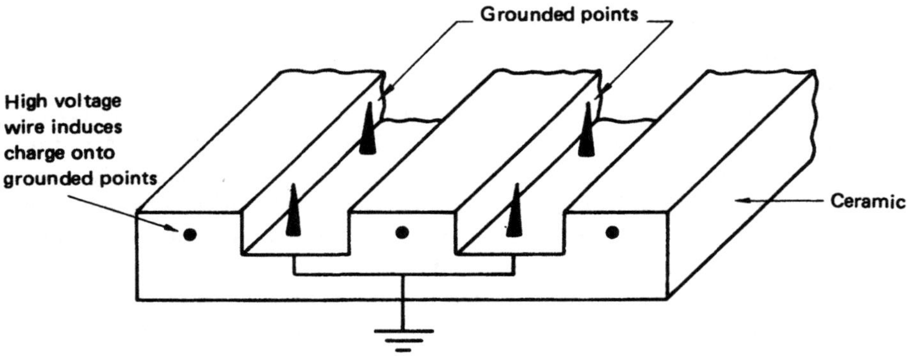

FIGURE 4.6. Encapsulated corona discharger.

different types of eliminator, but results from one particular installation with moving belt is shown in Table 4.5. In this example the radioactive bar used in combination with a passive discharger was the most effective. It is quite common for AC dischargers to overreact and deposit a small charge of the opposite polarity after neutralization. At lower fields the passive discharger would be less effective. However, in this case passive bars did produce a significant decrease in charge. Whatever neutralizer system is chosen the positioning is important. The passive and AC corona source both depend critically on the spacing from the charge to be neutralized. All the dischargers rely on the attraction between the unwanted charge and charges of the opposite polarity in the plasma created by the discharger. Therefore, it is important that the attractive force of the unwanted charge is at its maximum at the point of neutralization.

In practice this means that dischargers should be mounted far from

TABLE 4.5. Field Meter Readings Adjacent to Moving Web, Downstream of Antistatic System[a]

Type of eliminator	Field meter reading (V/m)
Standard tinsel bar	8×10^5
Tinsel bar with only 25% of bristles	6×10^5
Tinsel bar with only 10% of bristles	4×10^5
Passive pin bar	3×10^5
0.015-in. diameter wire	5×10^5
0.0048-in. diameter wire	2.5×10^5
Radioactive combination bar	5×10^4
Electrically powered alternating source	1×10^5

[a] Antistatic bar 1–1.5 cm from web.

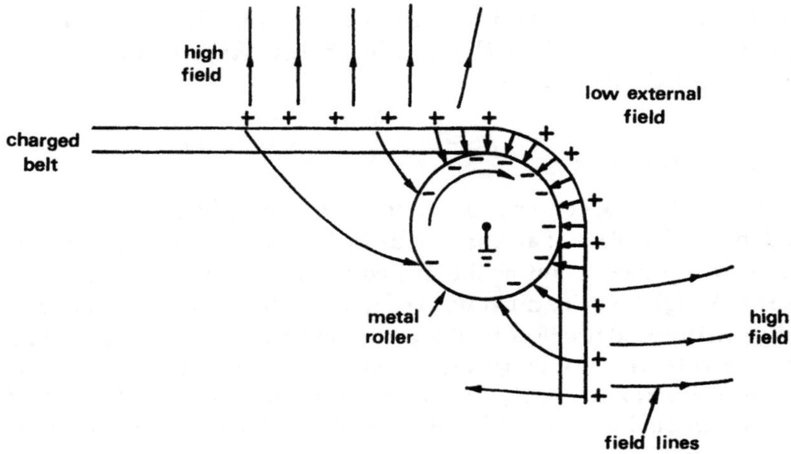

FIGURE 4.7. Field as charged belt passes over grounded roller.

any other grounded objects. The problem is illustrated in Figure 4.7. When the conveyor belt passes over the grounded roller, charge is induced in the roller forming an image charge. Since the unwanted charge is then paired its effect is small and the discharger ions are not attracted to the belt. In fact, it is even possible that if the charge source is not adequately screened ions will be attracted to the grounded roller and hence to the roller, adding charge in alternating stripes to the belt.

When very low charge levels are vital, a closed-loop controlled neutralizer system can be used (Figure 4.8). The charge on the surface after neutralization is detected with a field meter and the result fed back in order to control the output of a dual-bar discharger. If the belt is positive, the negative voltage of the dual bar will be increased slightly and for negative charges there will be an increase in the positive voltage. It has been found that the device is most effective if the field meter monitors the field for several seconds and averages before adjusting the

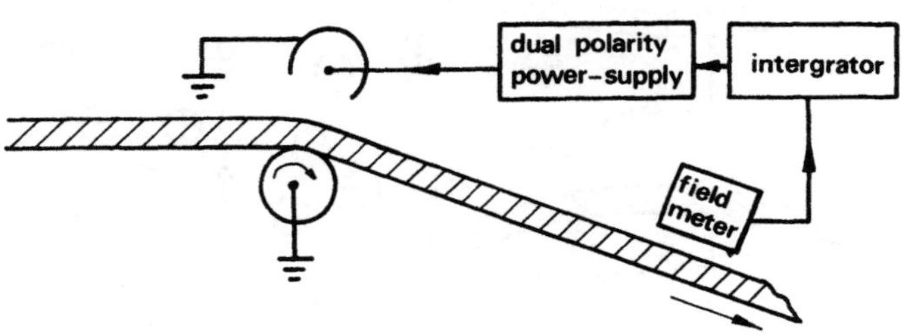

FIGURE 4.8. Closed-loop control system.

potentials. This prevents hunting but does mean that neutralization is only possible for gradual changes in charge. Rapidly varying charge densities are difficult to control.

4.4.4.3. Jet Ion Generators

When the charge to be eliminated is inaccessible dischargers may be used in which a blast of air carries the ions. The plasma may be produced either by a corona point as illustrated in Figure 4.9 or by a radioactive source. A high velocity must be used in order to carry the ions out of the orifice without attraction to the walls, and the ion current is found to increase with velocity up to supersonic speeds. However, the efficiency of these devices is nearly an order-of-magnitude less than the type without an air blast, possibly because the turbulence generated by the blast encourages recombination of the ions.

4.5. ELECTRIC SHOCK

The hazards of electric shock are detailed in the International Electrotechnical Commission Publication 479. The hazard of an electric shock is not the magnitude of the voltage but of the current. Thus a high-voltage power supply giving 50 kV and having a resistor of $10^9\ \Omega$ on the output can only give a current of 50 μA. This is safer than a supply of 500 V with no output resistor which, in flowing directly through a man of resistance $10^5\ \Omega$, gives a current of 5 mA. If the man has wet hands—reducing his resistance further—a higher current still will flow.

The current that a man can withstand depends on his health and the duration of the flow of current through him. For direct current 0.5 mA gives little reaction while at 13–16 mA it is still possible to let go. But at higher currents muscular spasms occur so that if a wire is held it is not

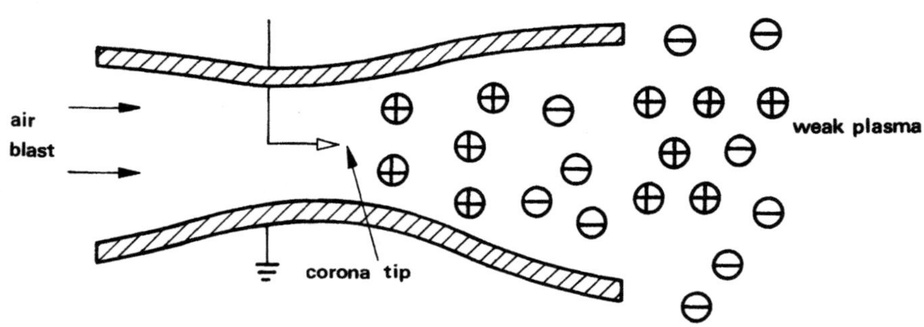

FIGURE 4.9. Jet ion generator.

possible to release it, although up to 1 A can be withstood for less than 20 ms. Electrostatic sparks are of less than 1-ms duration and the peak current is generally less than 1 A.

Less work has been carried out for short pulses than for longer-duration contact because electrostatic sparks are seldom directly dangerous to operators. A spark of the order of 4 mJ is the minimum that will be detected by a human being and discharges above 10 J are dangerous.

Any detectable spark may cause an accident if it makes the operator jump, so that for example, he falls from a ladder or drops tools. It is more common for an electrostatic spark to cause an accident indirectly in this way than directly.

Case Histories—Electrostatic Ignition Sources

Loading of Anthraquinone Powder

Static electricity ignited anthraquinone powder that was being loaded from paper bags into a hopper. A section of the metal hopper chute had been replaced with plastic and measurement of resistance to ground showed that this had insulated the hopper from ground. The result was an eventual buildup of static electricity which, on discharge, ignited the powder. The operator was immediately enveloped with flames due to the surface layer of powder that had accumulated on his overalls during the emptying of the bags and he later died.

The fire spread to other bags of anthraquinone powder stacked to a height of about 1 m on pallets on the floor. Firemen had difficulty fighting the fire since they had to climb over these bags; this displaced powder from the bags where the ends had been burnt away. A momentary flash-over occurred which slightly singed four firemen.

There was severe damage by fire to electric cables, plastic pipes, and the hopper compartment. About 700 kg of anthraquinone powder was destroyed.

Chocolate Crumb Silo

Estimated loss was £165,000. The premises consisted of a two-storied silo room and an interconnected single-storied store. The silo room had a concrete floor and brick walls to first-floor level with lightweight alloy wall and roof construction. The roof truss and columns were of unprotected steel. Intake and extract ducting was sited at the head of the silos. Lighting was by electric wiring in metal conduit to fluorescent strip lighting and heating by steam pipes.

During loading, chocolate and sugar particles in suspension in the silo were ignited. The resulting dust explosion blew off the hatch and metal ducting from the top of the chocolate silo and caused severe damage to the lightweight metal walls and roof of the silo room. The glazed roof of the interconnected store was destroyed, and the brick walls at roof level split horizontally. An opening at ground-floor level between the silo room and the store room allowed the blast to vent into the store, causing severe damage.

Three employees were cut by flying glass and taken to the hospital for treatment. Following the initial explosion, several small smouldering fires developed in nearby sacking

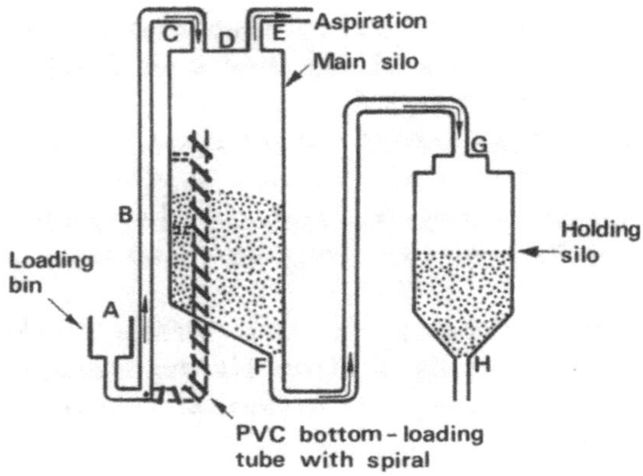

FIGURE 4.10. Schematic diagram of chocolate crumb silo.

and were quickly controlled by the Fire Brigade. A previous small dust explosion had occurred in the same silo in March 1974 causing no structural damage.

The explosion occurred following a change to bottom filling by means of a rigid PVC pipe (Figure 4.10). Several other installations identical except for the pipe, were running without problem. A model was set up in which chocolate powder was pumped through PVC pipe in the same way as in the silo. Field meters showed intense electrostatic activity in the model and an ignition occurred. No problems were observed with a metal pipe. Examination of the pipe showed small black pin pricks indicating that a spark had occurred between the inside of the pipe and the grounded metal strap wound around it.

A very high charge had been able to build up inside the pipe on the plastic surface and this eventually caused an electrostatic spark to the grounding wire which ignited the chocolate crumb.

Methanol Vapor Fire in a Silo

Dry organic material was transferred from fiberdrums with plastic liners through a metal funnel into a silo which had previously contained a methanol-moist product. Before the transfer, the silo was flushed with nitrogen blown in from the bottom during the entire transfer operation. Despite this, a deflagration took place in the funnel and two operators sustained burns from the resulting flash fire.

It appears that the relatively large opening used for filling the silo prevented an effective displacement of the air by nitrogen. Methanol vapors emanating from the methanol-wet product still sticking to the walls of the silo were ignited by a discharge of static electricity caused by friction between the plastic liner of the drum and the metal funnel.

In the future, the solvent-wet product will be transferred into metal containers only after effective inerting and without the use of plastic liners.

Near-Miss in a Cyclone Dust Separator

Originally, the inner cone of a cyclone dust separator was clamped with its own flange between the upper flanges of the cyclone. Since the cyclone was grounded, the inner cone was also grounded. A check revealed that the flanges of the cyclone had to be reinforced to withstand a possible internal dust explosion. When the modified cyclone was reassembled, the inner cone was no longer electrically bonded to the main structure and was being isolated by a nonconductive gasket.

Shortly after startup, sparking was observed on the outside of the cyclone between the flange of the inner cone and the flanges of the cyclone due to an electrostatic charge generated by the dust.

This is an example of a near-miss that illustrates the hazard of an ungrounded conductive element of an comparatively high capacity in an environment where static electricity can be generated. It also emphasizes the dangers that can arise when plant modifications are made without proper hazard analysis, even when the modifications are intended to improve the safety of the plant.

Explosions of a Dust Air Mixture

A sorbose powder was being transferred from an aluminum-lined cardboard drum into a reaction vessel containing acetone. A flame flashed out of the grounded metal funnel. Although the fire was put out immediately by the automatic dry-powder extinguishing system, four operators were badly burned and one of them died.

The fire was caused by the explosion of the dust/air mixture initiated by a discharge of static electricity in the charging funnel (sparking between the aluminum liner of the drum and the metal funnel or between a metal poker and the funnel). Air containing flammable solvent vapors or gases may be ignited by a much lower energy than pure dust suspension in air.

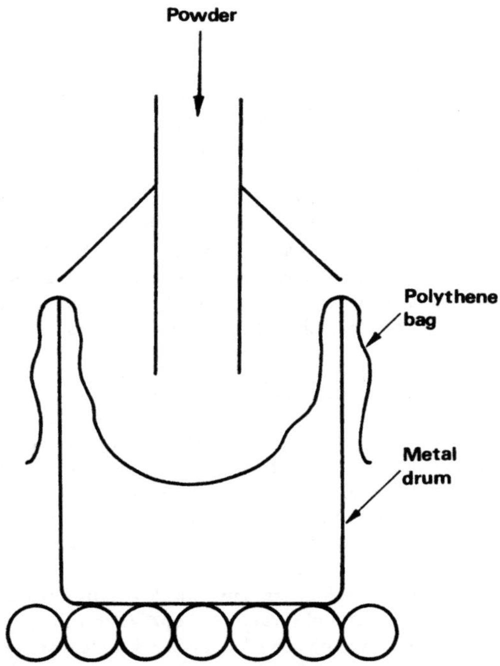

FIGURE 4.11. Loading of fine powder into polyethylene lined bins.

Polyethylene Lined Drums

A number of fires occurred in which luckily no one was injured. Grounding checks showed that the grounding of plant, men, and drums was adequate. It was found that although the powder was conducting and the hopper and chute were grounded, the movement of the powder down the chute applied a charge of 6×10^{-8} C/kg. (This is typical of manual-filling operations.)

Providing that the grounding rod was in the bag, the charge decayed to earth instantaneously when the powder settled in the bag. However, careful observation of the operation showed that frequently when a new bin was put in place, the full bin with the grounding rod was simply pushed out of the way. The butterfly valve was reopened for the new bin and the grounding rod added only after the powder in the full bag had been shaken down and the top sealed.

The powder in the polyethylene bag of the new bin thus represented a charged isolated conductor. In fact, since no one was close enough to be injured it is likely that the spark occurred not as the grounding rod was added, but as the weight of the powder in the bag became sufficient for the bag which was normally placed half in the drum (as illustrated in Figure 4.11) to fall to the bottom.

This case history illustrates three points: (a) Even slow movement of conducting powders in grounded pipes can produce sufficient charge for a spark if the charge is allowed to accumulate on an isolated conductor. (A total charge of greater than 10^{-7} C is required.) (b) Isolated conductors occur easily particularly in manual operations. (c) It is unsatisfactory to rely on the workforce to routinely apply grounded straps or rods if the operations may be more quickly completed without them.

Polythene Powder Explosion in Silo

A large silo filled with polythene powder exploded. Radio detection indicated that electrostatic sparks were occurring in the silo but it proved impossible to identify the source of ignition more precisely, since the production silo had only a single manhole in the roof through which instrumentation could be lowered. It is possible that the source of ignition was a spark from the highly charged settled pile of powder, but other sources of ignition from a highly charged insulating powder can also be postulated. Three solutions to this problem can be proposed. (a) Static eliminators can be inserted to discharge the incoming power. (b) Grounding rods can be inserted to encourage self-discharge of the powder. (c) The system can be inerted. In the case in question inertion was chosen because safety criteria for charge levels in storage bins are not well-defined.

Grain Elevator, U.S.

A powdered antibiotic was delivered to a feed mill. The antibiotic was pumped up a 4-in. plastic pipe running inside the man lift. A ground wire ran alongside the pipe. Unfortunately, a break in the pipe meant antibiotic had been pumped onto the floor of the man lift and a reclaiming company was called in to deal with the problem.

They vacuumed the powder off the floor into their truck then pumped it back into the 4-in plastic pipe which had been repaired.

The people operating the reclaiming had shut down the system several times because they heard "popping" sounds and observed "hot sparks" on the plastic pipe. The foreman then went into the basement and observed static sparks. He was in the process of shutting down when the explosion occurred. There were three fatalities and damage was estimated at $1,000,000.

4.6. SAFETY IN ELECTROSTATIC POWDER—COATING SYSTEMS

Electrostatic powder coating is a process in which an electrostatic ignition source and a flammable dust cloud are likely to be present at the same time. In this process powder usually less than 50 μm in diameter is blown into a cloud from a powder spray gun. At the tip of the gun there is a sharp point or edge to which between 50 and 100 kV is applied so that a corona discharge is created. The corona discharge causes the air to ionize and the ions are attached to the particles so that they become charged. The charged particles are then attracted to the grounded workpiece to which they adhere by electrostatic forces until the coating is fused in an oven to form a continuous paint film. Oversprayed powder is collected and recycled. The high-voltage source that provides the potential on the gun may be current-limited since only between 20 and 200 μA is required for effective coating. However, the process still creates highly charged powder and also free ions that will charge any ungrounded objects.

In fact, it is extremely difficult to ensure that all workpieces are always grounded since the jigs holding the workpieces tend to get coated with a thick paint layer as they are successively coated and passed through the oven.

It is also possible to draw a high-current corona that can degenerate into a spark from a small protrusion on a grounded workpiece, which then sets up a very high field as the protrusion passes in front of the high-voltage spray gun. However, in a conventional spray booth the cloud is only at explosive concentrations in the main spray cloud and this is not confined. Sparks from the workpiece are unlikely to ignite the cloud. If when a spark does occur in a dust-cloud region and the powder ignites, it is self-extinguishing because the volume at optimum concentrations is small and unconfined.

However, if a booth fire does occur, hot particles are created and may be carried into the recovery systems. It is here that major problems can occur because the dust concentration will be high in the areas where the dust is collected and these regions are also confined. The dust-recovery system normally consists of an air-extraction system followed by a cyclone and bag filter or sometimes by a bag filter alone.

The system can normally be designed with an air-extraction rate high enough to ensure that the concentration of dust in the ducting is below the minimum explosive concentration. However, at some stage in the collection system explosive concentrations are bound to exist. Since booth fires, although rare, are difficult to eliminate entirely and the powder is extremely highly charged (usually up to 10^{-3} C/kg—considerably higher than is produced naturally,) it is particularly important that the dust-recovery system is safely situated and adequately protected.

The oversprayed powder is to some extent a waste product and in many cases it is sufficient simply to ensure that any region where dust may reach explosive concentrations is adequately vented directly to the outside.

The most common source of a dust explosion in a powder-coating system is in fact hot particles from a booth fire, caused by an electrostatic spark in that area. It is therefore possible to protect the system by means of fire-detection systems that rapidly shut down the system and close valves to prevent the fire spreading to the recovery unit. Optical or heat detectors may be used in the booth but it is necessary to ensure that the sensors do not cease to operate because they become coated with layers of powder.

Although dust explosions are an obvious hazard in powder-coating plants, the fire and explosion hazard is still considerably smaller than when conventional solvent-based paints are used because the ignition energy of the dust is higher and the regions where the dust concentrations are hazardous are considerably more limited.

5

Powder Properties and Their Measurement

5.1. INTRODUCTION

The explosion of a powder or dust can be characterized by the ease with which it ignites and by the violence of the fire or explosion. Although the two are often related it is not necessarily the case that a material that is easy to ignite burns violently. Both the ignitability and the effects of ignition may differ according to whether the material is dispersed in a cloud or is more densely packed in a layer or heap.

A dust fire or explosion is essentially a rapid oxidation of the particle surface; it therefore depends both on the reactivity of the material and the surface area available. There is a strong dependence of ignition properties on particle size with fine particles more easily ignited and giving more intense explosions.

The water content of the dust may also be a significant factor because evaporation of water can take up some of the heat of reaction and prevent an explosion from proceeding so rapidly. Dust fires and explosions can be considerably affected by small quantities of an additive or impurity in the dust and it is important that these are known.

The behavior of a dust explosion depends not only on the reactivity of the individual particles but also on the way the heat is transmitted between particles. Ignitability and the explosive strength are therefore a function of dust concentration. The ignition of a powder or dust in cloud or layer form is a complicated process, and an extensive series of tests have been developed over the years to evaluate the ignition properties of materials in a way that can be applied to industrial plants.

All the tests are normally carried out on a fraction of powder that passes through a 75-μm sieve because fine particles separate easily in an industrial process and tend to be more explosive than coarser materials.

In order to evaluate a dust-explosion hazard it is necessary to know not only the concentration of dust present and the ease with which it ignites, but also the likely size distribution of the dust and the characteristics of the resultant explosion of the worst possible concentration.

The following measurements are required to completely characterize the hazard of a system:

1. *General Tests:* (a) particle size; (b) particle shape; (c) water content of particles; (d) composition of particles and their reactions under any conditions that may occur in a plant; (e) effect of possible additives and impurities; (f) powder resistivity.

2. *Ignitability*: (a) ability to ignite with a high-energy source; (b) minimum ignition energy; (c) minimum explosive concentration; (d) minimum ignition temperature of a cloud; (e) minimum ignition temperature of a layer; (f) temperature at which a powder deposit can self-heat; (g) maximum permissible oxygen concentration to prevent ignition.

3. *Results of ignition*: (a) maximum explosion presure; (b) maximum rate of pressure rise.

5.2. GENERAL PROPERTIES—PARTICLE SIZE AND SHAPE

It has been stressed throughout the previous chapters that the behavior of a dust depends on its size and shape as well as on its composition. The dispersal, flow, electrostatic charging, ignitability, and the rate and magnitude of the rise in pressure when a dust cloud explodes are all functions of particle size and shape. It is not possible to take measurements of any of these properties and apply them to the same material if the particle-size distribution is significantly different.

The particle-size distribution can be measured by a number of different methods, each of which measures a slightly different parameter. A review of measurement techniques is given in the book by Allen (1968). One of the simplest methods of sizing particles is to pass a sample through a series of sieves and weigh the quantity of material retained on each sieve. Dry sieving can normally be carried out down to 56 μm but the separation of dust sizes much below this requires air assistance or wet sieving.

Sieves tend to retain a percentage of particles that are of smaller diameter than the sieve mesh but that adhere to the mesh or to other particles. Nonspherical particles tend to pass through the mesh size equivalent to their smallest dimension. These two effects give rather broad cut points to the different fractions and, on significantly nonspherical particles, it may appear that particles are smaller than they really are.

Particles can also be sized either manually or automatically from photographs taken by an electron or optical microscope. Here one of the main difficulties, apart from the tedium of manual counting, is the selection of a representative small sample from a bulk powder. This is covered in detail by Allen.

Automatic techniques can considerably facilitate the particle-counting stage. A quantimet displays the photograph on a television screen. The picture is scanned electronically and the area of the particles measured. The equipment is computor controlled but can be interactive so that, for example, agglomerates of powder can be identified as separate particles by the operator. A shape factor as well as the size distribution can be calculated by the computer. The method is, however, only as accurate as the initial sample and if the particles are non-symmetric with a preferred direction of orientation the two-dimensional picture analyzed will not give a true size distribution.

One of the most commonly used methods of particle-size analysis is the Coulter counter. In this instrument a small sample of powder is suspended in liquid. Unless the powder is soluble in water this is normally water containing a salt to increase its electrical conductivity and wetting agents to ensure that the particles are well dispersed in the liquid. The liquid containing the suspended particles is drawn through a small orifice and passed between two electrodes.

When no particles are between the electrodes a steady current is measured. The passage of a particle between the electrodes causes a pulse to be registered. The pulse size is a measure of the change of resistance of the path between the electrodes which is a function of the volume of the particle. The pulses are counted by a pulse-height analyzer which gives the number of particles of each radius. In this case, the radius is that of the sphere with an equivalent volume regardless of particle shape. The machine can also be programmed to give distributions by volume, i.e., the total volume or weight of particles of each radius.

The Coulter counter uses only a small sample of powder so an effective method of choosing a representative sample must again be devised. Some powders tend to agglomerate and a microscope check is advisable to ensure that the additives included to ensure dispersal are adequate.

Sedimentometry is a sizing technique which measures the settling rate of powder either dispersed in air or in a liquid. The drag force on a particle settling through a medium of viscosity η, is given by

$$F = 3\pi\eta\, dv$$

where v is the particle velocity and d is the diameter. For a nonspherical particle, d is the Stokes' diameter, which is not necessarily the same as the diameter of a sphere of equivalent volume measured by the Coulter counter. The Stokes' diameter is the diameter of the sphere having the same density and terminal velocity as the sample.

Particles can also be classified by various air classifiers. For example, a cyclone classifier will separate coarse particles from fine ones because the heavy particles are thrown out to the edge of the conical cyclone whereas the fine particles are entrained with the air and forced up the center of the cyclone in a countercurrent.

Powder particles in a dust cloud can be sized by optical methods. The attenuation and scattering of light from a particle is a function of particle size. In modern instruments a laser beam provides the light and illuminates dust passing through a small focus point. The scattered light is analyzed and with the aid of computer program the size distribution of the dust can be found.

The different particle sizing methods each give a different particle radius for nonspherical particles and exact comparisons cannot be made between dusts measured by the different methods. In all methods the number or mass of particles in each given size range is found. These size distributions are often expressed in terms of the median size, i.e., the size at which 50% by weight of the particles lie below that size and 50% by weight above. The median size of a distribution gives only a rough guide to the likely behavior of a cloud because a powder containing a relatively small percentage by weight of fine particles may be considerably more explosive than a material of lower median size which lacks the fine particles.

5.3. DUST RESISTIVITY

The electrical resistivity of a powder or dust is relevant to both electrostatic and electrical sparks. From an electrostatic point of view, the more conducting the powder the less likely it is to store charge, and the less prone it is to causing static problems. However, a conducting powder is more likely to cause a short circuit in the vicinity of electrical equipment and in this context it is more hazardous. American classification divides dusts into three groups E, F, and G according to resistivity as follows:

(i) Group E—dusts with resistivity lower than $1\ \Omega \cdot m$ or which break down when subjected to 1000 V in the standard test cell;

(ii) Group F—dusts with resistivity between 1 and $10^6\ \Omega \cdot m$ which do not break down at 1000 V in the test cell;

(iii) Group G—dusts with resistivity greater than $10^6\ \Omega \cdot m$ which do not break down.

If a group-E dust lies on a circuit between live parts, a short circuit with high heat release can be expected. Group-F and -G dusts should not cause a short circuit, but that does not mean that the dust will not affect the operation of the instrument. Even the most insulating classificiation G contains dusts that are still highly conducting, from an electrostatic point of view where dusts are considered to be conductors incapable of storing charge until the resistivity exceeds $10^9\ \Omega \cdot$ m. The AC resistivity may be measured in a standard test cell described in American Standard S.P. 12.10. The circuit used to measure resistivity is shown in Figure 5.1. The cell consists of rectangular cross-section electrodes 9.9×1.4 cm^2 mounted on a PTFE block. An alternating high voltage is supplied by a 100:1 transformer that can give up to 11.5 kV. The current flowing through the powder is measured with the ammeter. The resistivity is given by ρ.

$$\rho = \frac{kV}{I}$$

where k is the cell constant. This standard cell differs in a number of ways from the cell used for measurement of resistivity in evaluating the charge-decay time for electrostatic tests described in Chapter 3. The electrostatic measurement is usually concerned with insulating powders and high voltages. It is therefore much more important to prevent high-field effects and stray leakage currents by using guard rings. These are omitted from the test for electrical leakage through powders.

The biggest difference in the two tests is that one is using alternating current and the other direct current. When a potential is first applied across a powder between two electrodes there is an initial high current which falls to a steady value with a time constant RC, where R is the

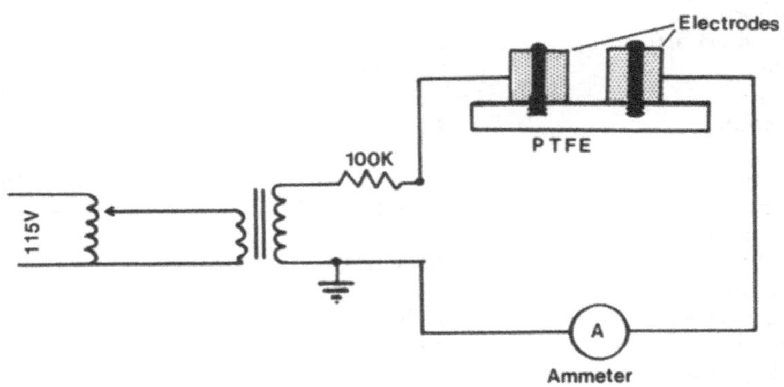

FIGURE 5.1. Alternating-current apparatus for measurement of dust resistivity.

resistance of the path between the electrodes and C is its capacitance. This high current is the displacement current and represents the energy absorbed as the molecules polarize in the field.

The value to which the current eventually falls is the steady DC leakage and is used for calculation of resistivity. If the alternating current is 50 Hz the sample remains at one polarity for a time of 0.01 s. Assuming the capacitance of the cell is 5 pF then any resistance exceeding 2×10^9 Ω will lead to a time constant for the powder reaching its steady level which is longer than the time available before the field reverses. A much higher conductivity will therefore be measured by the AC method. this is clearly illustrated in Table 5.1 which shows the resistivity of selected powders measured by the two methods. Both methods give a resistivity which is a strong function of voltage. Graphs for resistivity of sugar, cocoa, and flour are shown in Figures 5.2 and 5.3.

It can be seen that resistivity is not an absolute property of a powder and depends very strongly on moisture content and on the method used

TABLE 5.1. Powder Resistivity.

Material	Average resistivity, (Ω · m), applied voltage 1–10 kV	Minimum resistivity, applied voltage 1–10 kV
AC Measurement[a]		
Sugar	7.7×10^8	5.4×10^8
Sulfur	5.6×10^8	2.7×10^8
Cocoa	5.4×10^8	3.7×10^8
Grain dust	4.1×10^8	2.2×10^8
Cornstarch	2.4×10^8	2.1×10^8
Flour	2.3×10^8	2.0×10^8
Baking powder	1.9×10^8	1.5×10^8
Activated carbon	4.0×10^4	4.0×10^4
DC Measurment[b]		
Spray dried lactose	5.3×10^{13}	1.2×10^{13}
Lactose	9.7×10^9	4.8×10^9
Sugar	3.0×10^{11}	1.7×10^{11}
Chocolate	4.2×10^{11}	1.3×10^{11}
Cocoa	—	1.1×10^{12}
Icing sugar	—	2.2×10^{13}
Barley dust	4.0×10^7	—
Dried barley dust 24 h at 100°C	2.0×10^{13}	6.4×10^{12}
Soya protein	—	1.6×10^{10}
Calcium caseinate	2.0×10^{13}	5.5×10^{12}
Aluminum	1.5×10^7	—
Carbon (charcoal)	2.0×10^6	—

[a]For average resistivity, applied voltage = 1–10 kV.
[b]For average resistivity, applied voltage = 400 V.

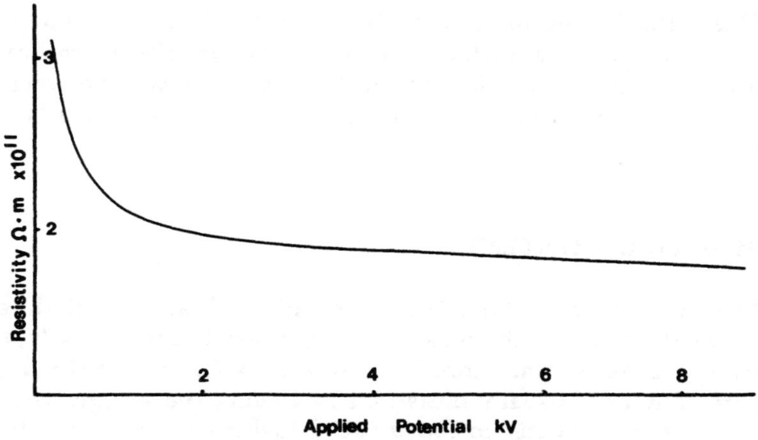

FIGURE 5.2. Resistivity of sugar—DC method.

for measurement. The test method chosen must depend on the reason for which the measurement is required. If the objective of the test is to evaluate the tendency of the dust to short-circuit mains-powered electrical equipment, then an AC test is required, whereas if the effect of an electrostatic potential is needed then a DC test at high voltage must be performed.

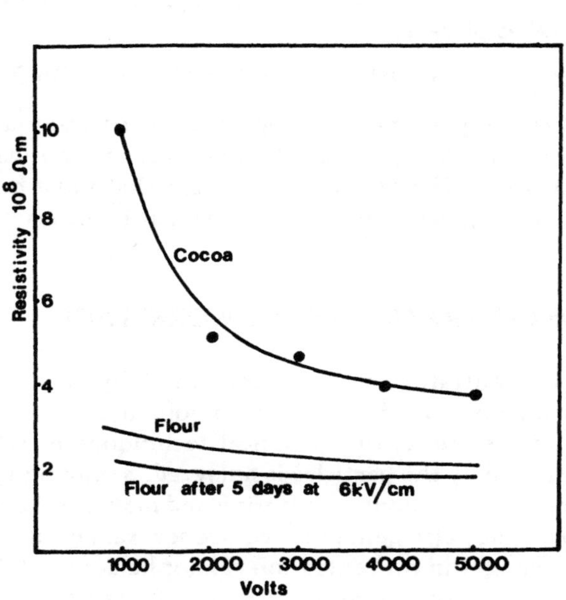

FIGURE 5.3. Resistivity of cocoa and flour—AC method.

One of the dangers of a powder that gets into electrical equipment is that it will carry a current which causes it to heat up. The AC test can be augmented by measuring the temperature of the powder between the electrodes. A rise in temperature of more than 10% is considered to be a potential hazard.

5.4. MOISTURE CONTENT

The moisture content of a dust can significantly affect both the ease with which it ignites and the results of the ignition. In addition, a dry dust is often less cohesive and more likely to form a flammable cloud. The moisture content is usually measured by finding the weight decrease when the sample is heated to above 100° C to drive off the water. If this temperature causes decomposition or oxidation, water may be extracted at a lower temperature in a vaccum oven.

This method is very simple and effective for materials that have a high moisture content of a few percent of the bulk weight, but it is not possible to completely dry off all the water condensed in surface recesses to measure extremely low levels of water content. In this case a chemical method may be used.

The most common chemical method involves the Karl Fischer reagent. This is iodine, sulfur dioxide, and pyridine in methanol; the reaction which takes place is

$$I_2 + SO_2 + H_2O + CH_3OH + 3py \rightarrow 2pyH^+I + pyHOSO_2OCH_3$$

(where py represents pyridine). The operation is basically a titration. When all the water is used up, iodine is in excess and can be detected visually or electrically. The quantity of reagent that must be added to produce excess iodine gives a measure of the quantity of water present.

5.5. DUST CONCENTRATION IN THE WORKPLACE

The dust concentration may be measured by direct sampling techniques in which powder from a known volume of air is separated from the air and measured, or by optical techniques in which light scattered or absorbed by the particles is analyzed. Sampling techniques are not applicable to continuous monitoring and may interfere with the normal dust flow. They also generally require several minutes sampling time so sudden changes in concentration cannot be detected. However, they are direct and it is possible to analyze the collected dust in detail for chemical nature, size, and shape.

Optical methods suffer from the problem that readings are not direct because the attenuation or scattering of the light is a complicated function of particle size. However, they do make measurements without disturbing the cloud.

5.5.1. Dust Cloud Sampling

Airborne particles tend to follow lines of an air stream but if the air encounters a surface the particle may strike the surface because it has inertia even though the air is deflected. Particles smaller than a certain size (depending on the air velocity) will be carried along with the air stream. This forms the basis of operation of an impactor sampler. In a cascade impactor, air is sampled and passed through a series of jets impacting onto sampling slides. The jets become progressively finer at each stage increasing the particle velocity. The coarsest particles are thus impacted onto the first stage with progressively smaller particles collecting at subsequent stages. A back-up filter may be employed to collect the smallest particles. This technique is not generally employed for high concentrations usually relevant to dust explosions but is useful when it is required to know the size of the dust as well as its concentration.

Potentially hazardous concentrations are usually measured by more-direct filter methods. A disk sampler consists of a vacuum pump that samples air at 45 liters/min for approximately 5 min. The air is passed thorugh a filter on which the dust collects. The quantity of dust collected is weighed and the air output measured by a rotameter. This gives the dust concentration per unit volume of air.

In any sampling technique involving suction an inaccuracy is introduced if the suction velocity exceeds the natural velocity of the particles, because different sizes of particles are accelerated by different amounts in the air stream created by the suction. Sampling in still air is therefore not a very accurate procedure except at slow sampling velocities. If the cloud contains any large particles, allowance must be made for their gravitational motion, for example, by sampling both vertically upwards and downwards.

Short-term sampling measurements may be made on a tape spot sampler. The air is normally sampled for between 1 and 20 s and the dust collected on a sticky tape. The dust is analyzed by means of a photo densitometer which measures the light scattered by the dust particles.

5.5.2. Optical Absorption

When particles pass through a beam of light they cause the light to scatter. Pulses of scattered light may be detected with the intensity depending on particle size, shape, and refractive index and the rate of pulses depending on particle concentration. This forms the basis of optical instruments using either laser or conventional light sources, which are used for measurement of particle size, distribution, and concentration.

Cloud density may be measured very simply by detecting the ratio between the light intensity entering the cloud f_0 to that emerging f.

$$\frac{f}{f_0} = e^{-\omega s l}$$

where s is the surface-area density, l is the path length, and ω the extinction coefficient. For particles greater than 1 μm, $\omega = 2$. To obtain absolute values from this equation it is necessary to ensure that no scattered light enters the detector. This means that the angle subtended by the detector entrance at the cloud must be very small ($\sim 0.2°$) and a reasonably intense light source is required to provide sufficient intensity for the detector.

Analysis of the scattered light to give particle concentration is more complicated and seldom used except when the method is also being used to measure particle-size distribution.

5.6. IGNITABILITY

Dusts can be classified according to the ease with which they can be ignited to give an indication of the probability of a fire or explosion hazard in an industrial plant. The classification of dusts is not as clear-cut as the classification of a gas. If the concentration of a gas mixture is known, a well-defined quantity of energy will cause it to ignite. A powder cloud has many more variables than a gas mixture.

The ignition of a dust cloud depends not only on the reactivity of the material but also on the surface area of dust available (i.e., the particle-size distribution) and the separation of the particles. Turbulence also has a pronounced effect. The ignition of a dust cloud and a dust layer is not necessarily the same process, and the ranking by ignitability that is found from cloud measurements will not necessarily apply to layers. The reactions that occur as a solid heats and ignition begins are far more complicated than for a gas and may include oxidation, decomposition, melting, and vaporization. The result is that the ignitability of a dust cannot be characterized by any one measurement, and tests must be carried out to cover every situation that may occur in practice.

The majority of internationally used tests have developed from tests devised by the United States Bureau of Mines and their figures for the ignitability of different materials are widely available. Brown (1951) reviewed the tests in use up to that time and a review is also given by Dorsett (1960). Until very recently modifications to the basic test methods have been limited in the interests of continuity of results from the early tests through to the present day. Generally a comparison between materials was felt to be of more value than a detailed knowledge of ignitability, and the tests have been used as a means of classifying dusts into groups in order to define the precautions which must be taken in handling that particular class of materials.

However, detailed work in Switzerland and Germany has shown that in a few cases the equipment that had been used for many years tends to classify some dusts wrongly. It has been found that tests made in a large spherical chamber give results that represent more closely the behavior of large-scale systems. As yet there is not universal agreement on a standard set of tests that are internationally accepted and both the older series of tests and the more-modern measurements are still in use.

5.6.1. Explosibility of a Dust Cloud

A powder may explode to give either a deflagration or a detonation. Detonating explosives require special procedures and there are normally special regulatory and registration laws. Some materials particularly susceptible to ignition which react rapidly at relatively low temperatures cannot be allowed to heat up and, for example, cannot be dried by application of heat.

With a new product it may be necessary to carry out preliminary screening tests to decide whether a material should be submitted to a specialist laboratory for detailed tests. This decision may be made on the basis of the following consideration:

(i) The chemical formula. Some chemical groups are particularly associated with explosive behavior such as organic peroxides and some nitrocompounds.

(ii) Oxygen balance. This is a measure of the oxygen content of the material readily available for reaction. It is commonly used as a measure of the potential explosive violence of materials and allows a comparison to be made between the material under test and materials of known explosive violence. The basic reaction that occurs is

$$C_xH_yO_z + (x + y/4 - z/2)O_2 \rightarrow x\,CO_2 + y/2H_2O$$

The oxygen balance is defined as

$$\frac{-1600\,(2x + y/2 - z)}{\text{molecular weight}}$$

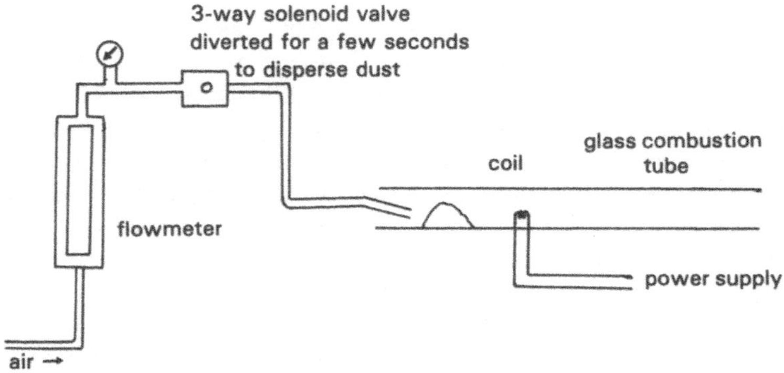

FIGURE 5.4. Horizontal apparatus. Dust is deposited in a lost heap in a tube 7.6 cm in diameter. The igniter is a hot coil. The powder is dispersed by a metered blast of air that blows it past the hot coil.

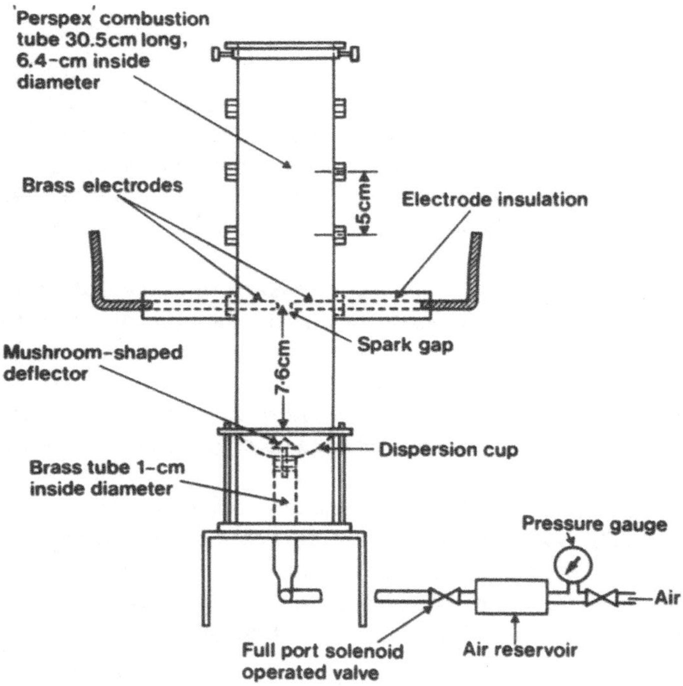

FIGURE 5.5. Hartman tube (1-liter volume). Dust is placed in the hemispherical cup and dispersed by air passing through a tube on the axis of the cup, which in turn is dispersed by the mushroom-shaped deflector. The dispersing air is released from a reservoir of fixed volume, the pressure of which can be varied. The ignition source is either a continuous arc or a heated coil and observations of the propogation of the dust cloud are made.

If a material has an oxygen balance more positive than −200 it is recommended for further tests, suitable for highly explosive materials.

(iii) If the considerations outlined above do not indicate highly explosive properties, then a small sample (of the order of 1 g) should be subjected to red heat or to a flame. If the reaction is not violent then the material may be tested according to the standard tests for normal flammable dusts described in subsequent sections of this chapter (Gibson, 1981).

5.6.2. Ignitability of a Dust Cloud

The ability of a dust to ignite is tested in England by the Fire Research Station, and in this country by the U.S. Bureau of Mines, in one of a number of forms of apparatus. All basically consist of a means of dispersing dust with different concentrations in the region of an ignition source. The dust dispersion is varied by altering the quantity of dust dispersed, the volume and pressure of the air used to disperse it, and the position of the dust cloud with respect to the ignition source.

Examples of the equipment used is shown in Figures 5.4, 5.5, and 5.6. The ignition source is generally either a heated wire ~30 J/s or a

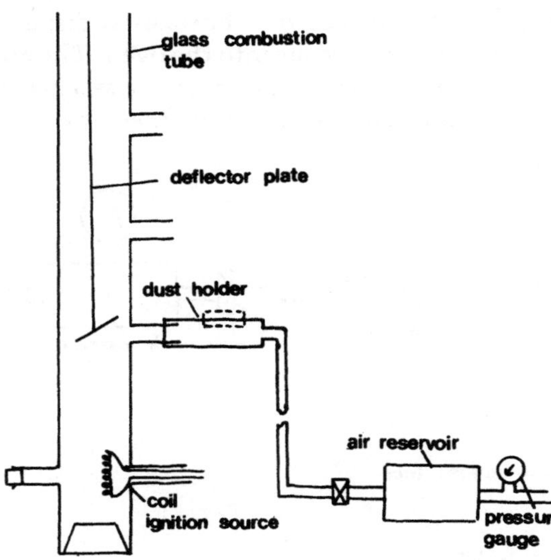

FIGURE 5.6. Inflammator. Dust is introduced from a dust holder via one of three side arms by a blast of air sufficient to blow all the dust against the deflector plate and form a cloud which falls on the igniting source. The source may be either a hot coil or a spark gap. Observation is made of whether the flames propogate away from the ignition source or toward it. This is particularly useful for dusts that are not readily dispersable.

continuous arc struck across electrodes with a 6-mm spacing. The arc is produced by a 5-kV alternating voltage from the secondary of a mains-driven transformer. If no ignition occurs, the dust is dried and sieved and the fraction that passes through a 75-μm sieve tested. A dust that still does not ignite is defined as nonignitable. However, this does not mean that the dust cloud cannot be ignited by any means. For example, PVC dust will not ignite with this test but will ignite in the tests carried out in Germany where a fusehead supplying 10,000 J is used as the ignition source. It is also possible that dusts which register as nonignitable will be able to be ignited with a moderate source at elevated temperatures. Some dusts in fact do ignite in the furnace apparatus described in Section 5.3.1 which do not ignite in any of the ignitability tests operated at room temperature.

5.7. THERMAL TESTS

5.7.1. Minimum Ignition Temperature of a Dust Cloud

The temperature at which a dust cloud will spontaneously ignite is measured in a furnace shown in Figure 5.7. The heated part of the furnace is an electrically heated quartz tube 0.037 m in diameter. A weighed sample of dust is placed in a horizontal dispersion tube and blown by a volume of compressed air into the oven. The air is previously stored in a small reservoir and the pressure in the reservoir can be varied to alter the dispersal conditions. Ignition is defined by the appearance of a

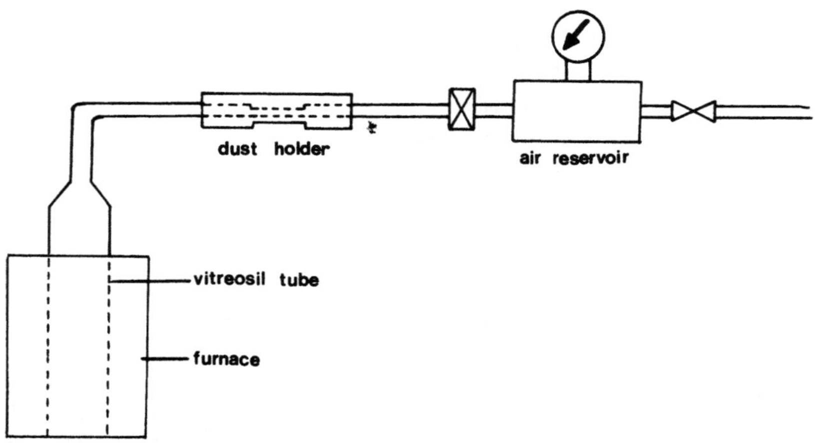

FIGURE 5.7. Godbert–Greenwald furnace.

flame at the base of the quartz tube. The tests are carried out until two temperatures are found that are 10° C apart and such that ignition occurs only at the higher temperature. The quantity of dust and dispersal conditions are then varied until the minimum ignition temperature is defined. In the test the dust passes fairly rapidly through the heated oven and a lower ignition temperature may be possible in a plant where the dust spends a prolonged time at elevated temperature. The minimum ignition energy of the dust cloud may be either higher or lower than the ignition temperature of the layer.

5.7.1.1 *Localized Ignition Source Test*

The ignition temperature of a dust cloud may also be measured using a localized ignition source such as a hot wire. The temperature required to ignite a dust cloud is generally significantly higher with a localized source than with the oven.

The localized ignition source test cannot be directly applied to any specific industrial source, but the relative classification of powders using a localized source is likely to be more relevant, for example to impact sparks than is the furnace test. The spread of combustion from a small source depends far too much on the nature of that source for the test to be of wide direct significance. Histograms for the relative ignition temperatures of dust layers and dust clouds measured in the Godbert–Greenwald furnace and with a localized ignition source are shown in Figures 5.8 and 5.9.

5.7.1.2 *Flame Spread*

Many dust explosions start as a result of fires in dust layers. The way in which a powder burns and fire spreads in a powder deposit is therefore extremely important. The fire properties are affected by particle size, packing density, and air movement. As yet there is no internationally agreed test, but test methods are basically similar. A layer of material approximately 10 cm long, 2 cm wide, and 2 cm deep is ignited at one end by a hot wire, flame, or chemical igniter. A controlled air flow may be passed across the surface.

The rate of propogation of burning and its nature are observed. The following classification may be made: (i) no fire spread—does not catch fire; (ii) brief ignition, extinguishing quickly; (iii) local combustion or smouldering—slight spread; (iv) glows red but no flame; (v) burns like a firework or slow steady burning with a flame; (vi) rapid combustion with flame or rapid decomposition without fire.

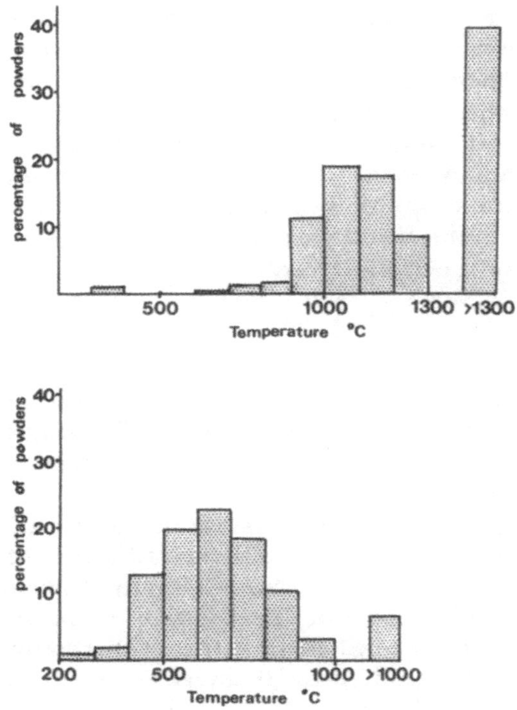

FIGURE 5.8. Ignition temperature of dust cloud: (a) hot coil; (b) oven.

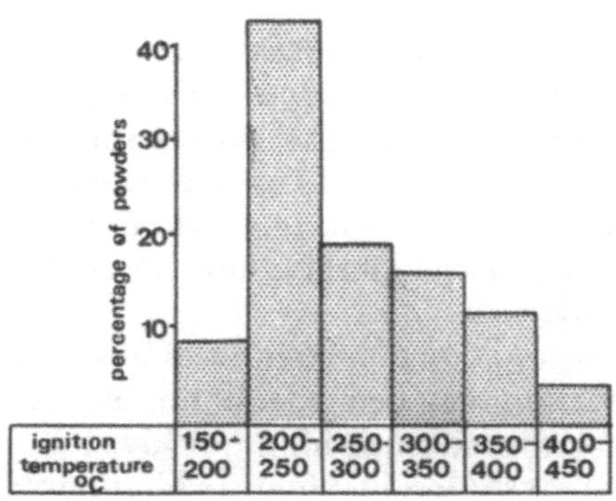

FIGURE 5.9. Ignition temperature of powder layers.

Decomposition products and any gases emitted during burning are also measured.

5.7.2. Ignition of a Dust Layer

A dust layer must exceed a certain minimum thickness before it can support combustion. This is determined by placing layers of dust of about 1-m^2 area on a surface at ambient temperature and igniting it with a wire heated to a dull red color. If the layer is too thin only the dust immediately touching the hot wires chars, and there is no propogation.

The minimum thickness at which dust continues to burn away from the heat source can be determined. The ignition temperature of a layer of dust may be observed by placing a layer of dust on a hot plate and observing the minimum plate temperatures at which the dust ignites or alternatively the temperature at which it chars, smoulders, or melts. The hot-plate temperature at which ignition occurs is a function of layer thickness because the powder acts as a thermal insulator, preventing the heat created by the dust oxidation from escaping. For example, the minimum ignition temperature of a sawdust layer decreases from 310° C at 1.2 cm thickness to 120°C at 180 cm thickness.

An arbitrary standard thickness of 0.5 mm was used in most early work on layer ignition temperatures. However, recent more sophisticated thermal tests have been developed to define the temperatures at which thermal runaway might be expected in a bulk sample. The tests identify the temperatures at which an exothermic chemical reaction occurs in the dust and take account of the heat loss through the sample. The temperature at which heat is created faster than it can escape can then be found.

The layer ignition test which is performed with thin powder layers and with a high temperature gradient across the sample can considerably overestimate the ignition temperature applicable to thicker layers or smaller temperature gradients. In some circumstances there may be a long induction time before the layer reacts. Corn starch will ignite at 500°C if the hot-plate test is carried out for 0.3 h, but the ignition temperature is reduced to 150°C if the temperature is maintained for 70 h.

The temperature to which a layer must be heated before it ignites depends not only on the nature of the dust but also on the rate at which it cools; it is therefore affected by the air flow. The flow of air across or through a layer may increase the minimum ignition temperature if its major effect is to increase cooling, but if it increases the access of oxygen to the sample the ignition temperature may decrease.

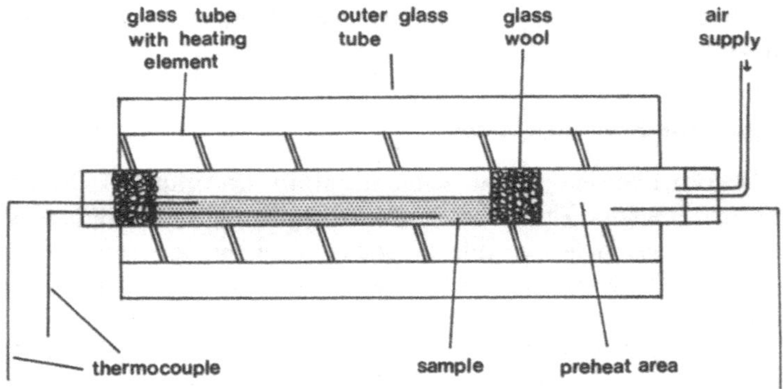

FIGURE 5.10. Horizontal tube with air flow.

Figure 5.10 shows an apparatus for investigating the ignition temperature in a horizontal tube with an air flow. The powder is held in a double-walled glass tube heated to the same temperature as the air. When the dust has reached the ignition temperature it begins to glow and shortly afterwards catches fire. In this apparatus the air is hot and therefore its most important purpose is to increase the likelihood of ignition (Grewer, 1971).

One approach to the problem of defining and measuring the different minimum ignition temperatures obtained under different conditions is to find the temperature at which an exothermic reaction begins and the energy released per gram of solid. The problem of the balance of heat loss and heat production can then be treated theoretically. The thermal analysis of chemical reactions has been developed into a very useful tool. The most common techniques used are DTA (differential thermal analysis), DSC (differential scanning calorimetry), and TGA (thermogravimetric analysis).

5.7.3. Differential Thermal Analysis (DTA)

In DTA the sample to be tested and a reference are heated in a metal block (Figure 5.11). The temperature of the sample and reference are monitored and the difference between the two temperatures is amplified and recorded as a function of time. The temperature of the metal block is increased linearly. Figure 5.12(a) and (b) shows the temperature of the two samples as a function of time and the equivalent DTA scan.

Normally small samples are used and the temperature can be scanned over a wide temperature range. The method gives the temperature at which an exothermic reaction begins, which may be lower than

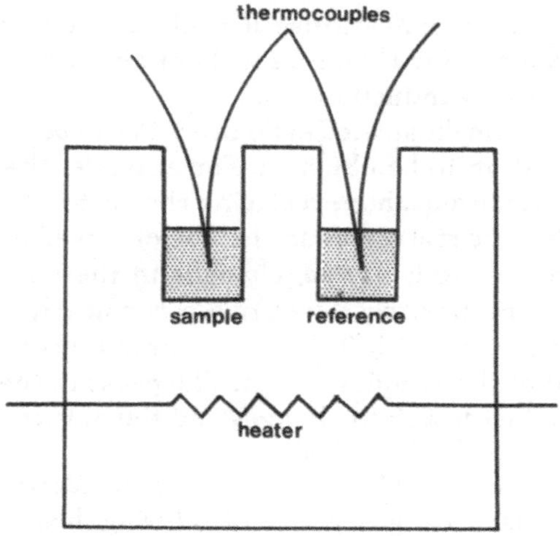

FIGURE 5.11. Differential thermal analysis.

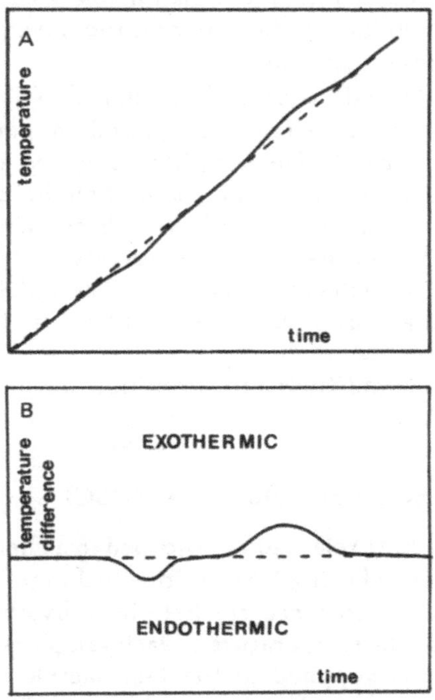

FIGURE 5.12a Temperature as function of time (dotted line is reference sample); (b) DTA scan.

the minimum ignition temperature if cooling is effective. DTA is a dynamic measurement and therefore will not give accurate results on a reaction that has a long induction time.

The simple system illustrated in Figure 5.11 can be modified to allow additional information to be obtained. For example, the system can be closed with an inert atmosphere replacing the air to prevent oxidation. Other less-exothermic reactions can then be observed in more detail.

If the scanning rate is varied, changes in the curves can help to identify reactions, particularly when two occur nearly simultaneously. DTA is often combined with TGA and a simultaneous record of the change in weight of the samples is kept. The peaks in the DTA scan can then be associated with weight changes and the nature of the chemical reaction occurring can be identified.

Evolved gas analysis (EGA) can also help identification of the individual reactions. The gas given off during heating is generally analyzed by gas chromatography or by infrared absorption spectroscopy. From the safety point of view it is also very important to know the nature and quantity of any gas given off. Gas evolution of any kind can be dangerous in a closed chamber but flammable gases are a particular hazard because of their low ignition energy, and the high risks associated with hybrid mixtures of gases and powders.

For some tests a modification of the standard TGA due to Boersma is used. The sample and reference are placed in pans on individual, thermally conducting bases. The samples are heated jointly by an oven, and the temperature of the samples and their holder is found. The thermocouples are fixed to the conducting bases which dominate the thermal capacity. This means that the temperature measurement can be more easily converted to the quantity of heat created, but the sensitivity is decreased and there may be some doubt about the actual sample temperature.

DTA methods and their interpretation have been reviewed by Pope and Judd (1977).

5.7.4. Differential Scanning Calorimetry (DSC)

The amount of heat released or absorbed in a chemical reaction can be more easily measured using DSC. In this technique, first proposed in 1964, the sample and reference are held in individual, independently heated, open dishes. The temperature of each sample is measured and the rate at which heat is supplied to the test sample is set so that the temperature of the two samples remains the same. The difference in the quantity of heat supplied, in the form of electrical heating energy applied to the two samples to maintain them at the same temperature, is monitored as a function of temperature. The measurement gives the temperature range at which a reaction occurs and the severity of the reaction, i.e., the quantity of heat created (or lost).

As with DTA the experiment can be extended to gain further insight into the nature of the reaction. For example, if DSC is carried out in a nitrogen atmosphere, oxidation is prevented and changes in the heat input indicate whether the exothermicity was due to oxidation. DSC is often carried out in conjunction with TGA. The main advantage of DSC over DTA is that it gives a direct measure of the quantity of heat created for each gram of reacting sample.

For accuracy, the heat losses from the sample and reference must be equal. It is therefore important that the emissivities of the sample and reference are the same. Where this is not the case, the samples are covered. DSC often spans a smaller temperature range than DTA but is more precise. Both DTA and DSC have the advantage that they require only very small sample sizes (typically only a few milligrams). With this quantity even very explosive materials can be tested. The time taken for the test is also short, with the full temperature range covered in 1–2 h. Although both these features make the methods ideal for initial examinations, both are unrealistic from an industrial point of view. A very small sample may not be representative of the bulk sample which is usually very heterogenous. In addition, the lowest heating rates of which the instruments are capable usually far exceed those in industrial situations.

5.7.5. Thermal Ignition and Self-Heating Tests

The temperature at which a reaction occurs, the induction time required before it reaches its maximum rate, and the quantity of heat released can be used to calculate the ignition temperature as discussed in Chapter 2. Often the situation is complicated both by the difficulty of calculating the heat loss from the sample and by the problem of identification of multiple reactions. For this reason a number of tests have been developed that attempt to simulate the conditions which may apply in practice to identify ignition temperatures.

If a sample is heated to a point at which it begins to self-heat a number of different temperatures can be defined: (i) The temperature in a given apparatus at which the powder sample temperature first exceeds the temperature of the oven, hot plate or other heat source; (ii) the minimum temperature of the heat source at which the sample temperature increases to ignition under isothermal conditions (i.e., the heat-source temperature is constant); (iii) the minimum temperature of the heat source at which the temperature increase in the sample proceeds to ignition under adiabatic conditions (i.e., the heat from the sample cannot escape). This would be the case in an industrial situation with a large powder sample but it is usually simulated in the laboratory by placing the sample in an oven and arranging for the oven temperature to follow the sample temperature to prevent heat loss from it.

These three temperatures are given various names in the literature but here the first will be referred to as the *minimum self*-heating temperature, the second as the *isothermal ignition temperature*, and the third as the *adiabatic ignition temperature*. The self-heating temperature and the isothermal ignition temperature are both a function of the heat loss from the sample, and therefore of the geometry of the apparatus, the size of the sample, and the temperature gradient across it. In layers these temperatures will also depend on convective heat loss from the sample, and the air flow through or across the sample will also be important.

A fourth parameter, which may usefully be measured, is the *ignition time* or the time for which the sample remains at the ignition temperature before ignition occurs. Again a distinction must be made between isothermal and adiabatic conditions. Measurement of the ignition time is complicated by the fact that it will be affected by the temperature gradient in the sample.

The self-heating temperature and isothermal and adiabatic ignition temperatures may differ quite considerably and it is important that the test method should be known before applying ignition temperatures to an industrial situation.

5.7.6. U.S. Bureau of Mines Method

The United States Bureau of Mines measures the self-ignition temperature using the apparatus shown in Figure 5.13. The apparatus is a double-walled Dewar to reduce heat loss from the samples. Five samples can be tested simultaneously and are hung in wire baskets in preheated air which allows the air to diffuse into the sample. The oven temperature is increased linearly and the temperature of each sample monitored. The temperature at which the sample deviates from the standard curve from a nonreactive material is taken to be the ignition temperature. This is effectively the self-heating temperature.

A similar apparatus is used in Germany but air is passed up through the samples on sintered plates. The test detects the onset of self-heating but it is usually dynamic so reactions with long induction times will be missed. It is not strictly adiabatic although heat loss is reduced by the Dewar. The size of the sample will affect the results as heat is lost

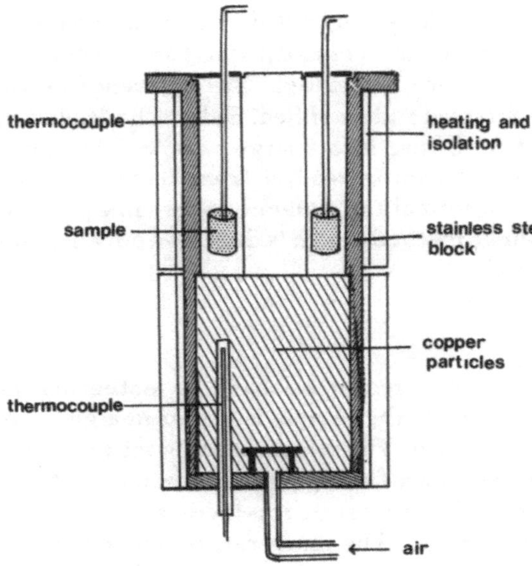

FIGURE 5.13. United States Bureau of Mines method for minimum ignition temperature.

through the sample once the sample temperature exceeds the oven temperature.

5.7.7. Giegy–Kuhner Technique

A similar method in which samples are heated in a metal block, but in which there is no air access, is known as the Geigy–Kuhner apparatus and is described by Zatka (1979). This is essentially a DTA technique. The temperature of the sample is increased at a heating rate of 2.5° C/min. When an exothermic reaction is observed the temperature is reduced by 10° C and a new sample is held at that temperature for as many hours an warrented by the industrial process. If the sample temperature exceeds the oven temperature, the temperature is reduced and the experiment repeated. This test gives the minimum self-heating temperature.

5.7.8. Sedex

Sensitive detector of exothermic processes (Sedex) is one of three techniques that have been developed recently for the specific analysis of thermal runaway reactions. (The others are Sikarex and accelerating rate calorimetry.)

In Sedex the sample is suspended in an oven heated by a hot gas stream. The temperature is increased linearly and the temperature of the oven and the sample are monitored. The difference between the two temperatures is detected and amplified. Sedex is basically similar to DTA with air allowed to diffuse into a large sample. The sample is heated uniformly in a gas stream instead of from below in a hot block. The measurement is isothermal and dynamic so the same problems exist as in the Bureau of Mines method which is different only in geometry and in the control methods.

5.7.9. Sikarex

Sikarex can be operated in adiabatic or isothermal mode. In the isothermal measurement the sample is placed in a glass container in a double-walled steel jacket. The temperature is set to a desired level and the temperature difference between the container and the center of the sample is measured. The temperature is increased slowly in steps using a fresh sample for each test. The temperature can be held steady at each temperature for as long as necessary to test for long induction times.

In the adiabatic mode the course of self-heating is examined by increasing the jacket temperature to follow the sample temperature. Sikarex equipment is fitted with a sophisticated control system so that the jacket is heated independently from the sample and there is no heat loss from the sample. This test gives the adiabatic ignition temperature. This represents the "worst case" giving the lowest temperature from which a reaction can accelerate to ignition if there is no heat loss from the sample.

5.7.10. Accelerating Rate Calorimeter (ARC)

In the accelerating rate calorimeter the sample is contained in a spherical bomb to which a transducer and thermocouple are attached (Figure 5.14). The oven is divided into different zones each with its own thermocouple. The bomb is separately heated by means of a radiant heater in a stepwise manner with a waiting period at each step. The temperature of the oven is maintained at the same temperature as that of the center of the bomb. The ARC carries out a heat-wait search, automatically measuring the rate of temperature rise on each occasion. When a predefined rate of temperature rise is reached, the oven follows the temperature of the sample until the reaction is complete (Figure 5.15). The reaction may safely come to completion, for example, if the reactants are used up before sufficient heat is produced for ignition or if an endothermic process such as melting intervenes. Alternatively it may proceed to thermal runaway and ignition.

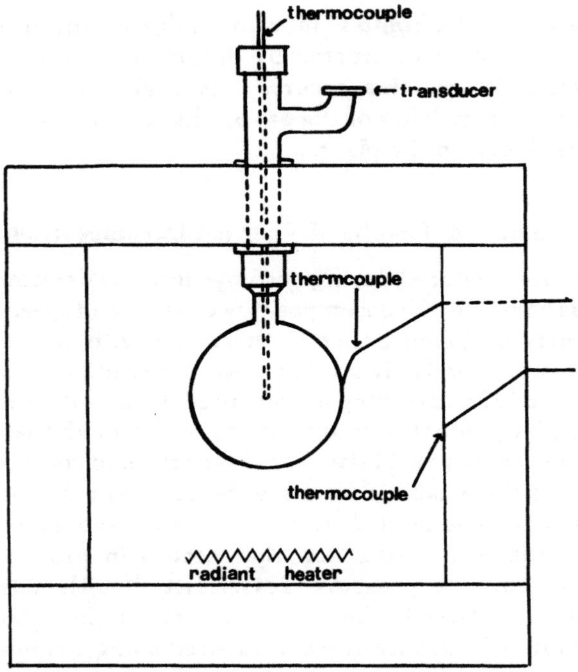

FIGURE 5.14. Accelerating rate calorimeter.

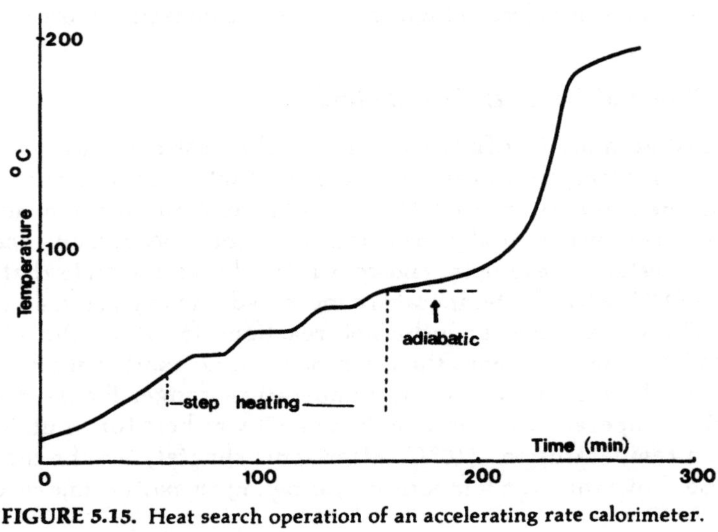

FIGURE 5.15. Heat search operation of an accelerating rate calorimeter.

The pressure in the bomb is monitored during the reaction. This gives information on gases created by decomposition or evaporation during the tests. Where this occurs, it is advisable to identify the composition and flammability of the gas produced as it may significantly affect the hazard level in the plant.

5.7.11. Measurement of Results of Thermal Decomposition

The pressure produced during decomposition is determined by the quantity of gas liberated, the decomposition energy, and speed. Pressure–time curves may be taken either isothermally with the temperature held constant or adiabatically so that once the reaction is initiated no further heat is added and cooling is prevented. In an industrial situation there is an initial gradual pressure increase but in the last phase of decomposition there is a rapid rise. For this portion of the curve there is little cooling and the adiabatic tests are the best representation of this situation. Tests must be carried out with a mass-to-volume ratio which adequately simulates the situation that occurs in practice. The experimental system is a pressure vessel fitted with a transducer. Thermocouples are placed in the sample and the whole system placed in an oven. The rate of pressure rise is measured using a transducer. The value found for this parameter during thermal decomposition cannot be directly related to the rate of pressure rise during the ignition of a dust cloud. During thermal decomposition there is often a slow pressure rise due to production of gas during the early stages of the reaction. The final rapid decomposition and explosion, therefore, often occurs in a system where the pressure is already high and the explosion is then more violent.

5.7.12. Choice of Thermal Test Method

To test each product fully with all available test methods is seldom feasible or necessary; however, it is important that all circumstances that prevail in the plant are covered. Hub (1980) gives a number of examples that illustrate how potentially hazardous situations escape identification in various tests. For example, Figure 5.16(a) shows a sample that was tested by DTA with the temperature increased linearly between 0 and 500° C. There were two endothermic reactions in which the sample absorbed heat compared with the reference, but no exothermic reaction was detected through the whole temperature range. Figure 5.16(b) shows the temperature of the sample when it was held for a number of hours at a temperature of 210° C. After approximately 3 h the temperature rose above the oven temperature, showing an exothermic reaction

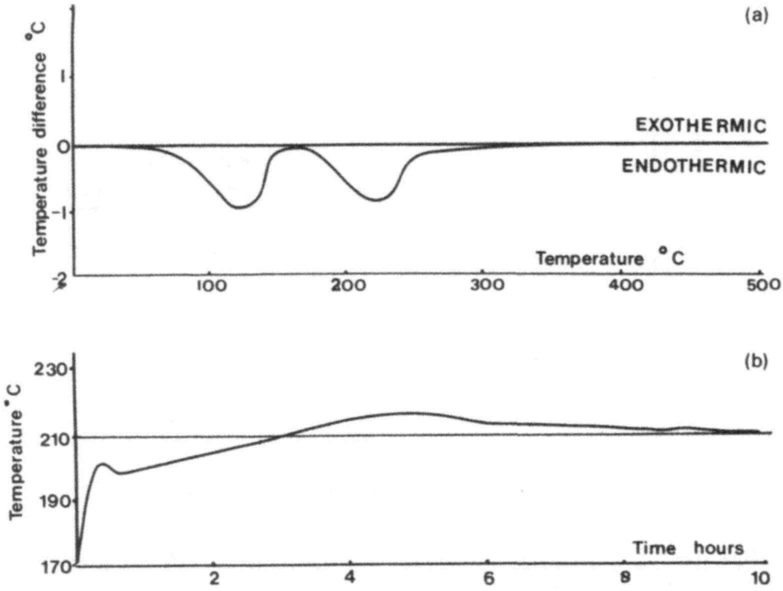

FIGURE 5.16. (a) DTA scan; (b) sample held at 210°C for several hours.

did occur. Since some time was required for the reaction to occur it would have been overlooked in any of the dynamic tests.

Problems can also occur when an endothermic reaction obscures the heat created in the tests. For example, Figure 5.17 shows a DTA exhibiting a strong endotherm but no exothermic reaction. In fact other tests showed that the heat absorbed by an evaporating solvent was masking the fact that heat was created. In a dry system a strongly exothermic reaction occurred. Although the tests allow the heats of reaction to be calculated it is not advisable to rely totally on calculation to ensure that cooling is sufficient to cope with experimental heat-production rate. The heating processes are frequently complicated with two or more different reactions involved at higher temperatures.

The test methods chosen may be static or dynamic, isothermal or adiabatic, with or without air flow. The most important criteria is that they mirror the conditions which the dust is likely to experience in the plant. Special tests may be defined if circumstances in the plant are not covered by standard tests; for example, the temperature of the surface which will ignite particles of falling dust may be measured.

Unfortunately some circumstances, such as very long storage times, may not be possible to simulate and the effect of time must then be found by calculation or computer simulation. Large volumes may be simulated

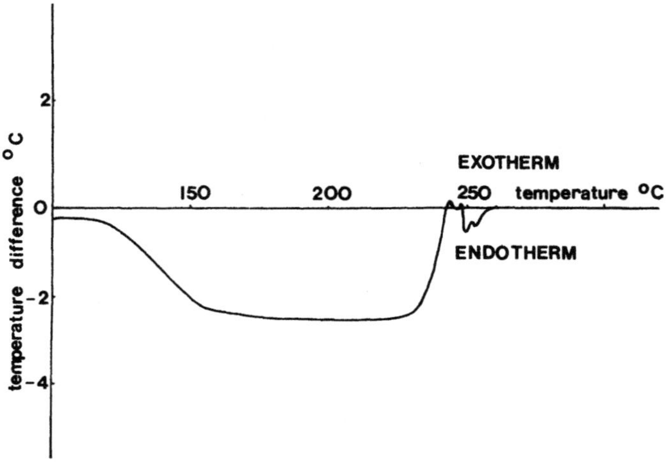

FIGURE 5.17. DTA showing endothermic reaction.

by adiabatic measurement preventing heat loss. This gives a good "worst-case" test but may overestimate the danger. Again in this situation calculations may be valuable providing the problems of multiple reaction are included. Examples of the use of data from the accelerating rate calorimeter to investigate the explosion hazard of diazalid are given by Townsend (1980).

5.8. MINIMUM IGNITION ENERGY

The basic principle of all the techniques used to study minimum ignition energy is to disperse the dust into a cloud which then passes between two electrodes. A spark of known energy is passed between the electrodes and the minimum ignition required to ignite the dust cloud is observed.

For many years the standard apparatus used throughout the world for this test was the Hartman tube, developed by the United State Bureau of Mines. All early published ignition energies were measured with this equipment. In the last 10 years it has been apparent that there are a number of design problems with this apparatus which cause it to overestimate the minimum ignition energy. Faults have been found with its size and geometry, the dust-dispersal method, the interpretation of the energy in the spark, and the timing of the spark. These problems will be discussed later.

Unfortunately, there has not yet been a universally agreed upon alternative system and a number of different arrangements, all giving slightly different results, are being operated in different laboratories. At present the only extensive lists of minimum ignition energies are those published by the United States Bureau of Mines and the British Fire Research Station. These both use the old Hartman vertical-tube system and overestimate energies by a factor of between 2 and 5.

5.8.1. The Hartman Apparatus

The Hartman vertical-tube apparatus, illustrated in Figure 5.5, is one of the pieces of equipment used in the general explosibility test. For minimum ignition energy measurements a spark from a capacitance is carried via a transformer to pass between the two electrodes. The energy of the spark is varied by changing the value of the capacitor. The dust is dispersed by means of a quantity of air stored in the reservoir which is suddenly released into the chamber by means of a solenoid valve. The dust concentration through which the spark is fired is varied by changing both the quantity of dust and the pressure of the dispersion air. Observations are made until the lowest possible energy at which the flame propogates through the cloud is found.

The circuit used to provide the spark is shown in Figure 5.18. The energy stored in the capacitor is equal to $\frac{1}{2}CV^2$. C is changed by adding more capacitors in parallel.

There are two major criticisms of the use of this equipment for minimum ignition energy. The most important is that it is assumed that the energy in the spark is the same as the total energy stored in the capacitor. This not only assumes that all the energy in the capacitor is dissipated—which is often *not* the case—but, more important, it ignores the energy lost in the transformer. The result is that the energy of the spark is actually considerably lower than the energy stored in the

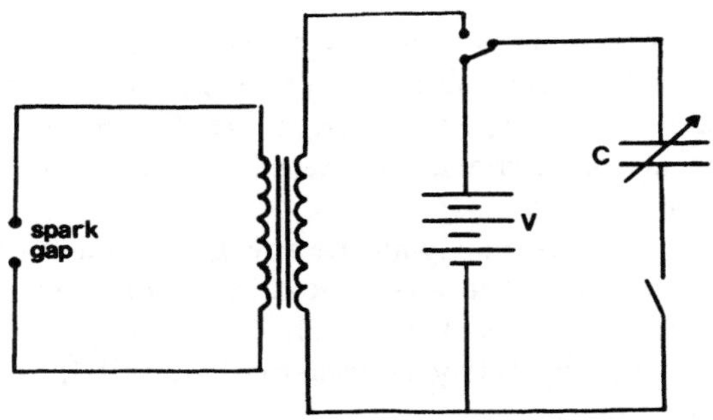

FIGURE 5.18. United States Bureau of Mines spark circuit.

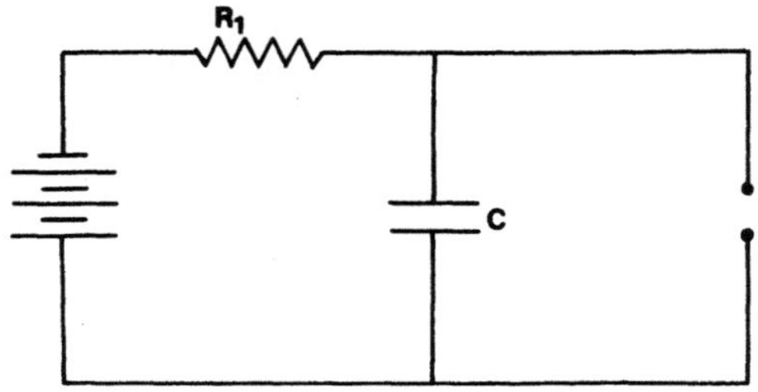

FIGURE 5.19. Direct capacitor discharge circuit.

capacitor and the measurement overestimates the minimum ignition energy.

The second criticism of the circuit is that it is not possible to control the timing of the spark accurately with respect to the dispersal of the dust cloud. Several measurements must therefore be made to allow for variations in the cloud concentration with time.

Although it is now agreed that the Hartman circuit overestimates the minimum ignition energy there is no universal agreement concerning which of the available alternative circuits should replace it. Unfortunately, the ease with which a spark ignites a dust cloud is not a function only of its energy; for example, long-duration sparks will ignite more easily than a short-duration spark of higher total energy, and sparks from large capacitors at low voltage ignite less easily than sparks of equivalent energy from a smaller capacitor at higher potential. A number of different circuits have been proposed. The simplest is a direct capacitive spark (Figure 5.19).

5.8.2. Direct Capacitive Spark Apparatus

A high-voltage DC power supply charges the capacitor C through the resistor R_1. When the voltage on the capacitor reaches the breakdown potential of the spark gap a spark occurs. The resistor sets the time taken by the capacitor to charge so there is minimal recharging for the duration of the spark.

When a spark occurs nearly all the energy stored in the capacitor is dissipated in the gap, and there is no loss in the discharging side of the circuitry other than in the spark gap itself. The energy of the spark in this circuit can therefore be fairly represented by $\frac{1}{2}CV^2$, where V is the potential breakdown.

The duration of the spark is given by C, where R_G is the resistance of the gap (of the order of a few hundred ohms). This varies between a few nanoseconds at energies of about one millijoule to a microsecond for energies of approximately one joule.

If it is required to know the energy more precisely, the current flowing through the spark and the voltage across the gap can be simultaneously measured using a high-speed oscilloscope. The method is described in Chapter 3. In this case it is necessary that the input resistance of the voltage-measuring circuit considerably exceeds R_1 otherwise the capacitor discharges through the voltage-measuring system faster than it charges.

In practice this normally means that the charging side of the circuit must be completely isolated from the discharging part of the circuit, and the discharge must be initiated after the capacitor charging is complete by an external source such as a small corona discharge. The most important point in favor of this circuit is that it directly represents the situation most common in industry—a spark from an isolated conductor—and it is possible to interpret directly in terms of the size of the capacitor at a given voltage that will give an incendive spark.

The critics of this circuit are worried about the relatively short duration of the spark for a given energy. It is not testing the worst possible case that can occur, although it is a good representation of the most common type of spark. It is also difficult to measure the spark energy accurately without very sophisticated equipment that can cope with the high spark speeds. The duration of the spark is extended by incorporating either a resistance R_L or an inductance in the discharge part of the circuit (Figure 5.20).

In the case of a resistor, the time constant of the spark is then given by $C(R_L + R_G)$. R_L is normally chosen to be 1–20 k Ω and hence R_G is negligible in comparison. R_L may be varied so the duration can be varied separately from the ignition energy.

The use of a resistor causes a significant amount of the energy from the capacitor to be dissipated in the resistor rather than in the spark gap.

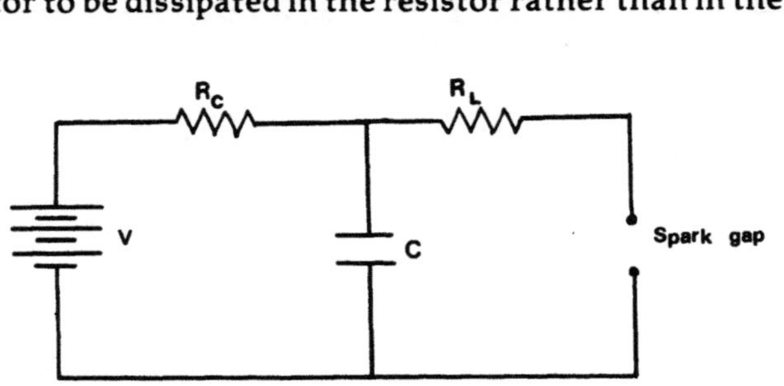

FIGURE 5.20. Circuit for long-duration capacitive spark.

For a direct current the energy loss would be I^2R_L, compared with I^2R_G dissipated in the spark gap (where I is the current flowing in the discharge circuit). In practice, the spark is a transitory signal and R_L and R_G must be considered to have shunt capacitances decreasing the energy loss. The result of the use of a resistor is that the energy dissipated in the spark gap can no longer be calculated and must be measured. However, measurement is easier because the spark has a long duration.

The charging side of the circuit must still be isolated during the measurement in order to allow the capacitor to charge effectively. If R_L is replaced by an inductance this also has the effect of extending the spark duration and no energy is lost. The spark length can therefore be altered at will without affecting spark energy. However, the spark is now oscillatory with characteristics depending on the value of inductor L as shown in Figure 5.21.

All the capacitance methods give lower minimum ignition energies than the U.S. Bureau of Mines' Hartman circuit. Lower energies still can be obtained by designing a circuit specifically to ignite dust efficiently. This approach has been taken by Eckhoff (1975). His circuit is shown in Figure 5.22. C is the discharge capacitor charged to the voltage V. The spark is initiated at the desired time by means of a trigger circuit. When the switch is closed a high-voltage pulse of 15 kV is induced in the

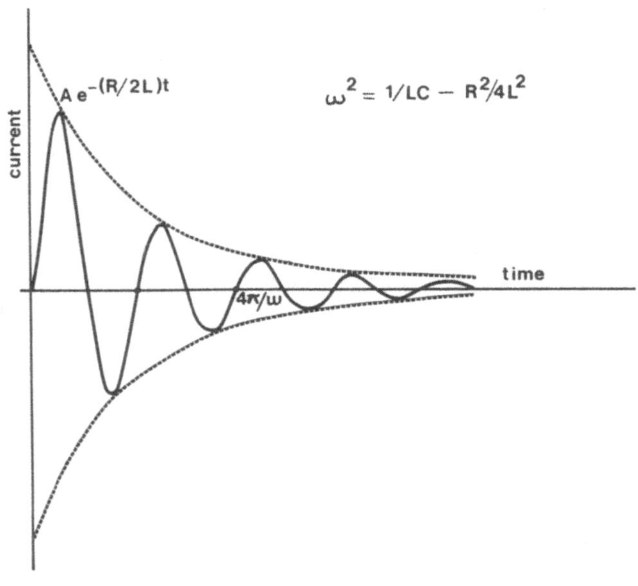

FIGURE 5.21. Inductive spark.

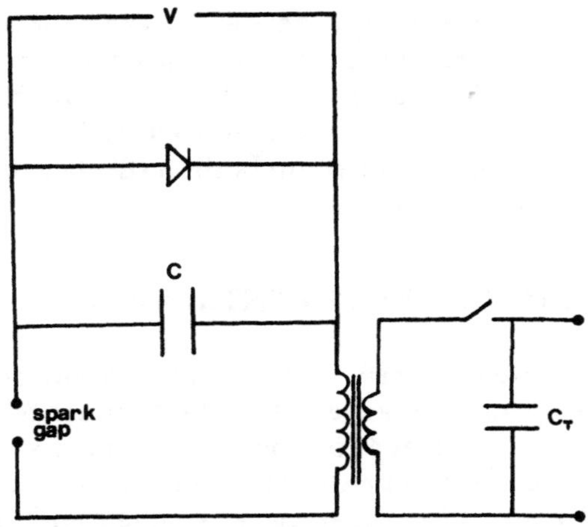

FIGURE 5.22. Spark circuit designed by Eckhoff (1975).

seondary transformer causing the spark gap to break down and discharge the main capacitor C. The energy stored in the secondary capacitor C_T is small compared with that in C, expect for low ignition energies below 10 mJ. Above this value, measurement showed that the energy in the spark was close to $\frac{1}{2}CV^2$.

The duration of a 10-mJ spark using Ekhoff's circuit is approximately 100 μs. This compares with 10 μs for the capacitor circuit with series resistance and approximately 50 ns for the direct capacitive circuit. Ignition energies of materials measured by the Eckhoff circuit, the Hartman circuit, and the direct capacitance circuit are given in Table 5.2. The sources of powder for the three measurements differed, therefore differences in ignition energy are to be expected; however, a distinct pattern does clearly emerge.

A single minimum ignition energy cannot as yet be defined with universal agreement, but for most practical purposes the objective of the

TABLE 5.2. Minimum Ignition Energies Measured by Different Techniques.

	Minimum ignition energy (mJ)		
Material	Eckhoff	Hartman	Capacitance
Wheat	20	50–60	—
Lycopodium	7	40	25
Sulfur	0.3	15	3–5
5-μm Aluminum	1	10–15	7.5

test is to draw dividing lines for powders which are unlikely to be ignited by electrostatic or other low-energy sources and those which are particularly sensitive. The point at which those lines must be drawn will be different according to the test method used, and the published data must be interpreted taking into account both the likely energy of ignition sources in the plant and the test method used.

5.9. MINIMUM EXPLOSIVE CONCENTRATION

A dust cloud must exceed a minimum concentration if a dust explosion is to propogate because the particles must be close enough together to exert some influence on each other. In the standard test for minimum ignition concentration the vertical tube Hartman apparatus is used. The quantity of dust dispersed is weighed and assumed to be uniformaly dispersed throughout the tube. The quantity of dust used is decreased until no explosion is observed in at least 10 tests. The criterion for indicating that an explosion has occurred is the visual observation of a flame propagating away from the ignition source. This is usually accompanied by a significant pressure rise in the tube, but the effect of the explosion decreases substantially at low dust concentrations and the pressure rise may be very small.

The Hartman apparatus creates a turbulent dust cloud of non-uniform concentration and gives no control over the part of the cloud in which the ignition source occurs. However, it gives a reasonable order of magnitude for the minimum ignition concentration. Normally when this parameter is used for the design of safety criteria (such as in defining an air throughput for a dust extraction system), a good safety margin is allowed and the approximations inherent in this test are not important.

Two other methods have been proposed for measurement of the dust concentration at the time of the spark. In the first a small pan is placed below the spark gap and the rate at which powder falls onto it is measured by weighing, this gives the quantity of powder through the gap at the time of the ignition. In the second method the dust concentration is measured by detecting the amount of light scattered by the dust particles. A small instrument suitable for this measurement has been described by Liebman, Conti, and Cashdollar (1977). Light is emitted by a light-emitting diode and detected by a photocell. LED and detector may be protected by quartz windows kept free from dust by means of an air jet. The instrument can then be used to monitor dust clouds in the factory as well as for detecting concentration in the Hartman tube.

Liebman *et al.* quote an expression for T, the fraction of light transmitted as follows:

$$T = \exp\left(\frac{-3Qcl}{2Dd}\right)$$

where Q is the extinction coefficient, c the mass concentration of dust, l the light path length, D the particle density, and d the particle-surface weighted mean diameter. Q/d is found by calibration against a known uniform dust concentration.

An optical technique obviously can only be used before the ignition source is initiated but is able to monitor the changes in dust concentration with time as the cloud is formed within the chamber.

5.10. MAXIMUM OXYGEN CONCENTRATION

A dust explosion requires not only an explosive concentration of powder and an ignition source but also a sufficient concentration of oxygen to support combustion. The maximum permissible oxygen concentration to prevent ignition depends on the composition of the other gases present. For example, different oxygen concentrations are permissable according to whether air is mixed with carbon dioxide or nitrogen for inerting.

Tests for the maximum permissible oxygen concentration are carried out either in the vertical-tube apparatus or in the furnace apparatus. The maximum permissible oxygen concentration at which no ignition takes place is significantly lower when the ignition source is a heated surface or furnace than when it is a spark. Values of 4–7% by volume are not uncommon in a heated environment although maximum oxygen concentrations are more usually 10–15% when the ignition is by spark or fusehead at ambient temperature. The maximum permissible

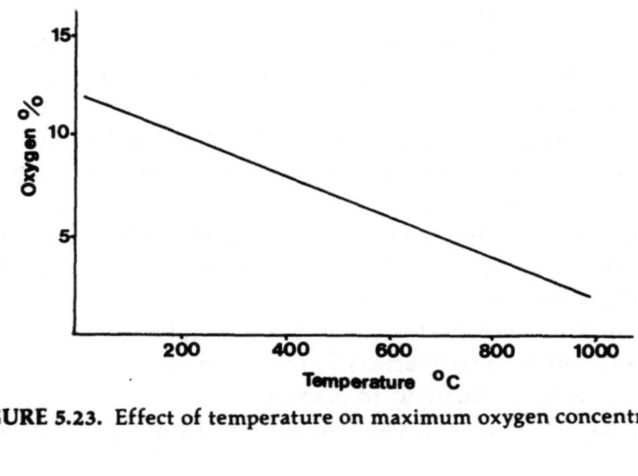

FIGURE 5.23. Effect of temperature on maximum oxygen concentration.

oxygen concentration is, in fact, a function of temperature; therefore furnace tests will lead to a much lower permissible level. Figure 5.23 shows the effect of temperature on the critical oxygen concentration for terephthalic acid dust in a nitrogen-inerted atmosphere.

Once again it is clear that the maximum permissible oxygen concentration is not an absolute figure which can be published to suit all industrial circumstances. It is necessary to know the test method and conditions in order to be able to apply figures to the industrial situation.

5.11. MAXIMUM RATE OF PRESSURE RISE; MAXIMUM EXPLOSIVE PRESSURE

The results of a dust ignition are characterized by the shape of the pressure pulse produced when the dust is ignited in a test apparatus. The pressure is recorded as a function of time and the maximum pressure and rate of pressure rise noted. When the test is scaled up to the industrial equipment, the maximum pressure defines the strength required to withstand the explosion and the maximum rate of pressure rise defines the speed with which venting or other protection must operate.

Early measurements of these parameters were made in the Hartman vertical-tube apparatus modified to accomodate an electronic transducer and strengthened to withstand the pressures produced. This is known as the Hartman bomb. The ignition source was either a continuous arc supplied by a high-voltage transformer, or a hot coil. The dust was dispersed by release of air from the reservoir (a higher air pressure was generally used then in ignition energy tests). The pressure curves were taken over a range of dust concentration because both the maximum rate and maximum pressure depend on concentration. The maxima do not normally occur at the same concentration.

This technique has come under even greater criticism for pressure measurements than for minimum ignition energies. The criticisms are summed up as follows:

(i) The Hartman bomb produces results that are difficult to scale up to a realistic size in order to estimate pressures and rates of pressure rise in an industrial scale system.

(ii) Since the heat source is continuous, it tends to ignite the dust cloud before it is fully formed; it therefore ignites a rather weak mixture and underestimates the results of the explosion.

(iii) Even if the method is being used simply to classify dust into three classes it will tend to misclassify some powders such as corn starch or sugar which may be difficult to disperse.

(iv) It tends to underestimate the effect of highly explosive dusts.

TABLE 5.3. Dust Classification

Dust explosion class	K_{st} (bar · m/s)	Example
St 0	0	Nonexplosive dust
St 1	>0–200	Grain dust
St 2	200–300	Organic pigments
St 3	>300	Fine metal dust

Bartknecht (1981) in Switzerland has carried out comprehensive tests investigating the effect of test-vessel size on the pressure and rate of pressure rise. It is found that a cubic law is valid for the maximum pressure rise in any vessel between 20 liters and 60 m³; i.e., if the maximum rate of pressure rise measured in any vessel is multiplied by the cube root of the volume a constant is obtained which is a characteristic of the dust:

$$V^{1/3}\left(\frac{dp}{dt}\right)_{max} = K_{st}$$

Dusts can therefore be classified by K_{st} value as illustrated in Table 5.3. The majority of dusts falls into the St 1 class. Some examples are given in Table 5.4.

The minimum-size vessel to which this law applies is 20 liters; below this limit, test vessles lead to a progressively lower K_{st} value. There is a tendency for the rate of pressure rise to reach a maximum in a small vessel, whereas it tends to continue to increase in a larger vessel. Small vessels therefore tend to badly underestimate the hazard of the very explosive St 3 dusts.

The standard apparatus for measuring K_{st} values is thus either a 20-liter sphere or a sphere of 1 m³. In both cases the dust is dispersed from a small container holding the dust sample and air under high pressure. This is released directly into the 1-m³ apparatus but the 20-liter

TABLE 5.4. Explosion Ranges of Combustible Dusts.

Nature of dust	P_{max} (bar)	K_{st} (bar · m/s)
Coal	7.8–8.0	60–97
Grain	8.6–9.3	98–112
Polyethylene	1.3–7.9	4–120
PVC	7.5–9.6	37–168
Epoxy	5.3–10.0	53–168
Flour	7.9–10.5	80–192
Organic pigment	6.5–10.7	28–344
Aluminum	6.5–13.0	16–1900

apparatus is evacuated so that the influx of air does not raise the pressure of the apparatus.

A number of different ignition sources are commonly used, including a capacitor discharge of a few tens of millijoules, a fusehead of 10 J, a pyrotechnical ignitor with an energy of 10,000 J, and a lump of smouldering dust. In many cases the K_{st} value is not changed by the nature and energy of the ignition source. However, some dusts such as PVC will only ignite with the high-energy source. A continuous arc (energy approximately 10 J), however, produces results that may be up to 60% lower than those observed when the other sources are used. This is probably a function of the fact that no delay can be used when the continuous arc is used and the position in the cloud at which ignition occurs can not be controlled.

A comparison of the 1-m^3 chamber and the Hartman apparatus has been carried out by Pineau, Giltaire, and Dangreaux (1976). Extracts from their results are given in Table 5.5.

The Hartman test gave lower maximum pressures by a factor of between 1.8 and 3. The maximum rate of pressure rise was very much higher in the large chamber for icing sugar, which did not disperse easily, and for aluminum. (The ignition source in the Hartman apparatus was a hot coil which does not ignite aluminum easily.) Many of the other results were comparable in the two pieces of equipment; however, it would be expected that the 1 m^3 would give lower rates of pressure rise by a factor of 10 because of the cubic law. The K_{st} value calculated from the Hartman apparatus is therefore an order-of-magnitude lower than that obtained in the 1-m^3 apparatus and Hartman results are not interpreted in this way.

The effect of the type and scale of test apparatus on the maximum explosive pressure has received less attention. Pineau et al. showed that the Hartman bomb gave significantly lower maximum pressures than the 1-m^3 sphere. This is likely to be partly a function of shape but also of the effect of the heat energy absorbed by the heavy steel walls in the smaller chamber.

A correlation between the maximum pressure obtained in a 1-m^3 chamber and a 20-liter chamber is shown in Figure 5.24.

As the maximum pressure increases, the value obtained becomes progressively lower in the 20-liter chamber. Scaling up to larger vessels is more difficult because it becomes progressively less likely that the dust concentration will be at the optimum throughout the chamber. The majority of tests in larger chambers have not been carried out in closed vessels and only results from vented chambers are available. Figure 5.25, for example, shows the maximum pressure in a vented chamber as a function of chamber size and vent area for corn flour.

TABLE 5.5. Comparison of Explosibility Measurements Made in a 1-m³ Sphere and the Hartmann Apparatus.[a]

Powder	Size (μm)				Maximum pressure (bars)		dP/dt (bars)	
	>125	>5–100	>100	Fine	Hartman	1-m³	Hartman	1 m³
Aluminum	—	8	—	—	4.85	12.4	118	2000
Wood floor	0.1	0.6	98.2	1.1	3.95	7.1	329	410
Dried milk	20.6	61.8	16.4	1.2	3.85	8.1	61	63
Epoxy	—	0.4	98.7	0.9	2.95	6.2	139	200
Aspirin	64.5	29.3	5.4	0.8[b]	3.0	9.2	293	277
Icing sugar	9.9	34.9	54.8	0.4[b]	3.23	6.2	91.5	215
Corn flour	2.5	18.3	78	1.2	3.6	6.6	71.5	95

[a] From Pineau *et al*. (1976).
[b] Probably agglomerating.

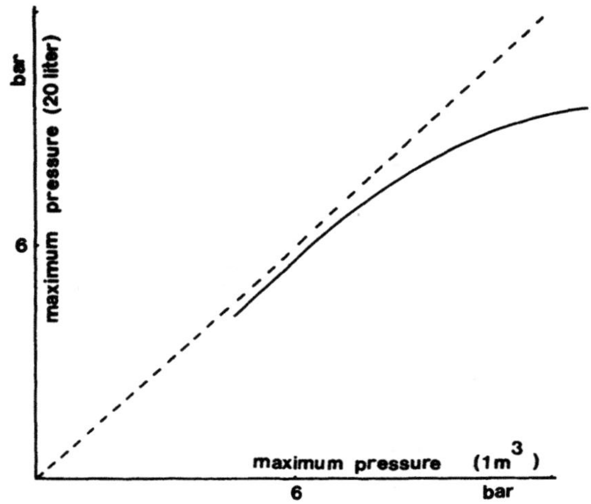

FIGURE 5.24. Maximum explosion pressure measured in 1-m^3 vessel and 20-liter vessel.

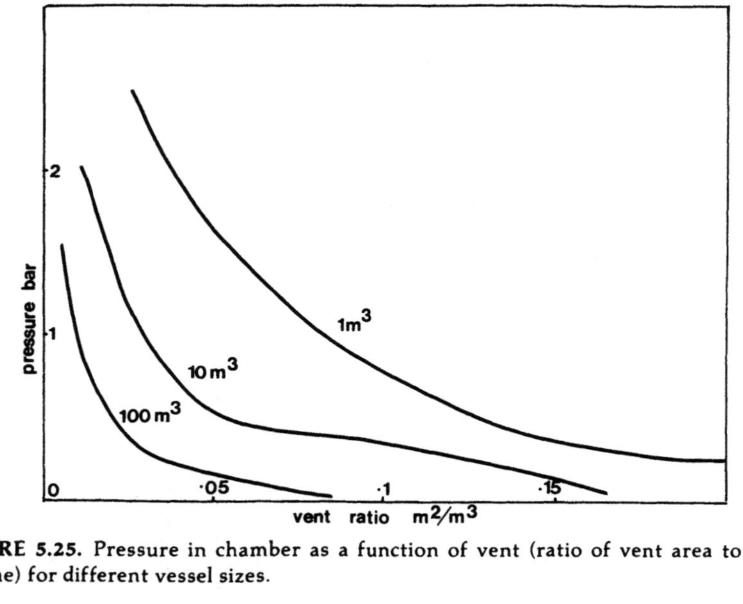

FIGURE 5.25. Pressure in chamber as a function of vent (ratio of vent area to vessel volume) for different vessel sizes.

It is found that the maximum pressure obtained in a vented chamber is, in fact, more directly related to the rate of pressure rise measured in a smaller, closed vessel than to the maximum pressure. Therefore, knowledge of the maximum explosion pressure is primarily required only when it is required to contain the explosion. Even then the transitory high pressure obtained during an ignition is difficult to correlate with strength calculations which normally give a vessel strength to a constant pressure.

Although the two standard parameters defined by the pressure curves are maximum explosion pressure and maximum rate of pressure rise, there is some evidence that the delay time between the time when the ignition source is fired and when the rate of rise reaches its maximum is related to the minimum ignition energy.

5.12. EFFECT OF PARTICLE PROPERTIES ON IGNITION PROPERTIES

The characteristics of the pressure pulse produced by an explosion and the minimum energy of the dust cloud are both a function of particle size. For most materials a graph of the maximum explosion pressure and rate of pressure rise as a function of median diameter are of the form shown in Figure 5.26. Both parameters remain fairly constant as size increases to 75 μm then falls rapidly so that by the time the particles reach 400 μm there is no explosion even with a high-energy source.

The weight concentrations, in g/m^3, at which the maximum pressure and rate of pressure rise occur, vary with particle size—being a direct function of the surface area of the particles and an inverse function of the distance between them. There is an upper limit to the rate of pressure rise that can be achieved because of the limited amount of oxygen in the chamber. Increasing the particle surface area beyond the limit where all the oxygen is used up has no effect.

Although it would appear that there is little change in parameters once the particle diameter is below 75 μm, published results show significant differences in explosion properties for different samples. Table 5.6 gives results for polyethylene taken from the German data.

Minimum ignition energy also decreases with particle size (Figures 5.27 and 5.28). These tests were carried out in the Hartman apparatus. Most powders become extremely difficult to ignite once the particle size exceeds a few hundred microns but there is some evidence that a very small proportion of fine particles less than 10 μm in diameter can considerably decrease the ignition energy of a large-particle-size sample.

The moisture content of a dust can considerably affect ignition

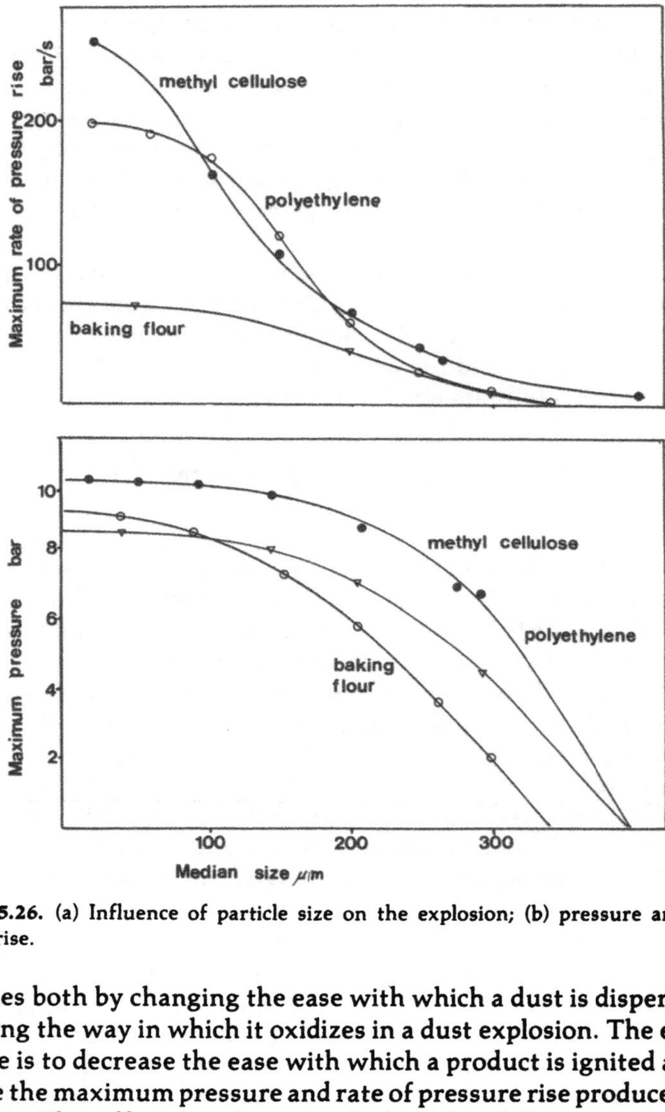

FIGURE 5.26. (a) Influence of particle size on the explosion; (b) pressure and rate of pressure rise.

properties both by changing the ease with which a dust is dispersed and by altering the way in which it oxidizes in a dust explosion. The effect of moisture is to decrease the ease with which a product is ignited and also decrease the maximum pressure and rate of pressure rise produced in the explosion. The effect has been studied in detail for grain products. Figures 5.29–5.32 show the effect of moisture content on ignition energy and temperature and on the results of the ignition.

The relative humidity and moisture content of a dust also affect its electrical resistivity. Thus nylon may have a resistivity of 10^{14} Ω · m in the dry state but be reduced to 10^{12} Ω · m after storage for 48 h in a humid atmosphere.

TABLE 5.6. Ignition Data for Polyethylene Samples of Different Sizes.

Particle distribution %					Median size	Minimum explosive concentration	Maximum explosive pressure	Maximum rate of pressure rise	Cloud temperature in furnace apparatus
⩽500	⩽125	⩽71	⩽32	⩽20 μm	(μm)	(g/m³)	(bar)	(bar/s)	(°C)
—	—	—	95	86	<10	(30)	8.0	156	420
—	—	—	94	80	13	15	7.6	82	360
—	—	—	70	37	24	15	7.6	74	400
—	—	—	94	21	24	15	7.6	54	410
—	—	—	65	39	25	15	7.4	62	450
—	97	73	—	—	52	30	8.8	122	460
—	—	70	6	—	62	15	8.5	131	440
—	—	56	16	—	65	30	7.4	62	470
—	79	47	22	14	76	15	8.2	110	450
—	62	42	21	14	90	(30)	8.4	157	420
—	70	28	—	—	105	30	8.7	109	440
—	56	—	—	—	111	30	8.5	80	460
—	50	21	—	—	126	30	7.8	65	510
—	44	21	—	—	135	60	7.8	58	470
—	36	10	—	—	150	125	7.4	54	480
—	28	7	—	—	160	(30)	6.8	36	480
—	10	6	—	—	162	60	7.4	58	470
100	25	9	—	—	163	60	6.9	48	490
—	24	7	—	—	163	(30)	5.5	13	480
100	25	6	—	—	170	60	7.3	49	500
100	18	—	—	—	173	250	7.1	38	470
99	1	—	—	—	182	—	No ignition		460
97	13	2	—	—	215	—	No ignition		460
98	—	—	—	—	230	—	No ignition		490
96	2	—	—	—	232	—	No ignition		460

(continued overleaf)

TABLE 5.6. (*Cont.*)

Particle distribution %					Median size	Minimum explosive concentration	Maximum explosive pressure	Maximum rate of pressure rise	Cloud temperature in furnace apparatus
⩽500	⩽125	⩽71	⩽32	⩽20 μm	(μm)	(g/m³)	(bar)	(bar/s)	(°C)
90	19	11	—	—	240	60	7.4	56	440
90	20	9	—	—	245	125	7.5	46	460
92	15	6	—	—	260	(30)	No ignition		450
91	13	4	—	—	270	(30)	6.9	39	450
95	1	—	—	—	280		No ignition		540
90	14	—	—	—	295	(30)	5.9	18	510
96	4	1	—	—	300	(30)	7.2	76	480
84	2	—	—	—	410	(30)	No ignition		590

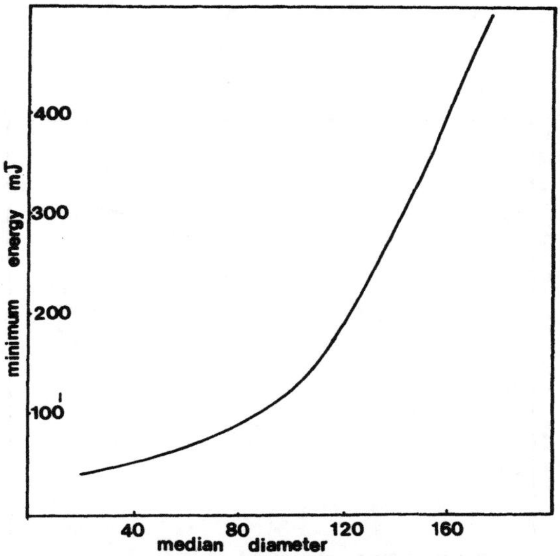

FIGURE 5.27. Minimum ignition energy of corn starch as a function of particle size.

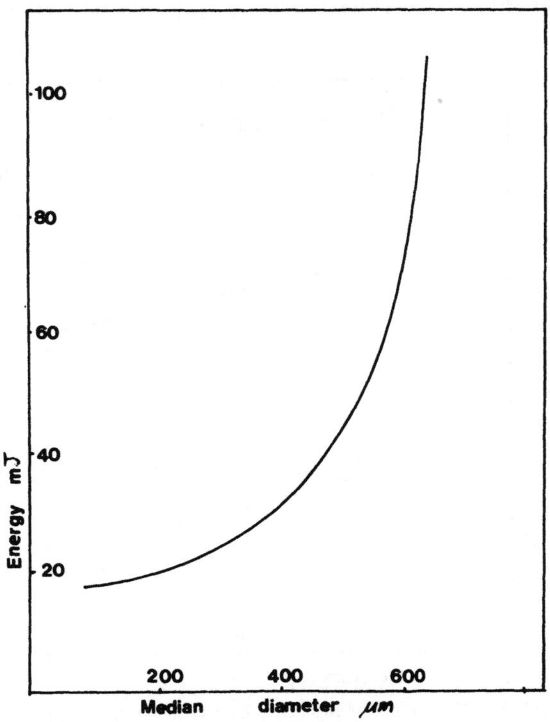

FIGURE 5.28. Minimum ignition energy as a function of particle size—high-density polyethylene.

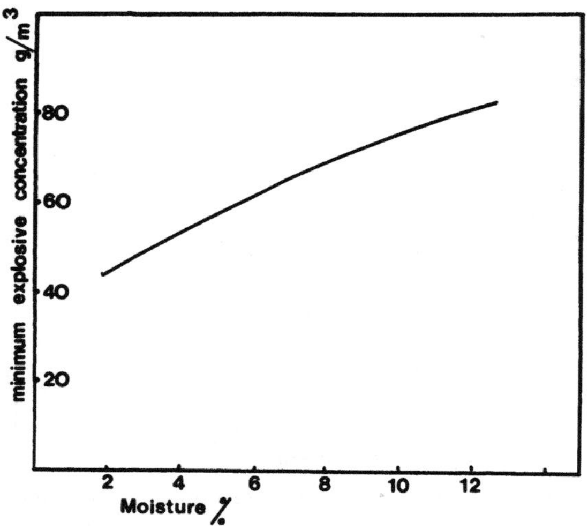

FIGURE 5.29. Effect of moisture on minimum explosive concentration of grain.

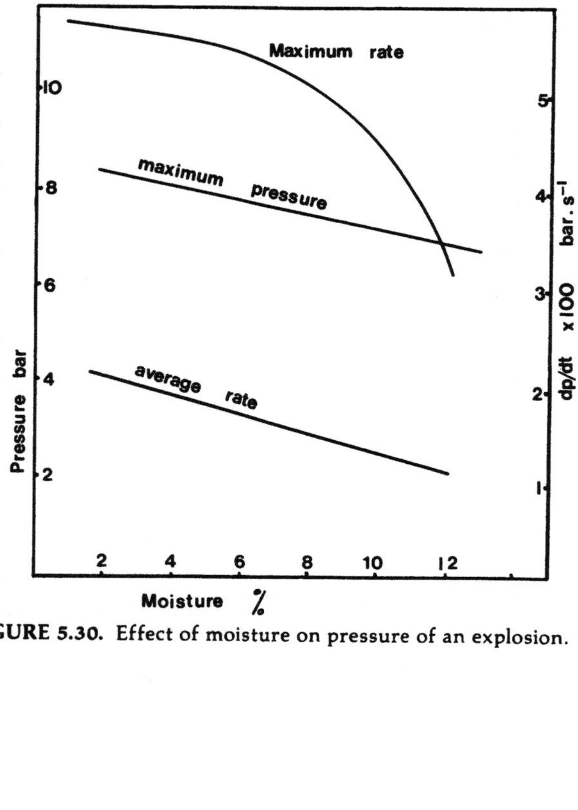

FIGURE 5.30. Effect of moisture on pressure of an explosion.

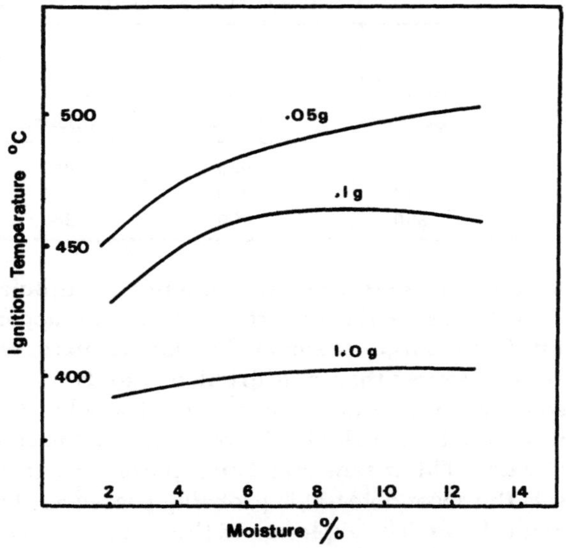

FIGURE 5.31. Effect of moisture on the ignition of temperature of grain.

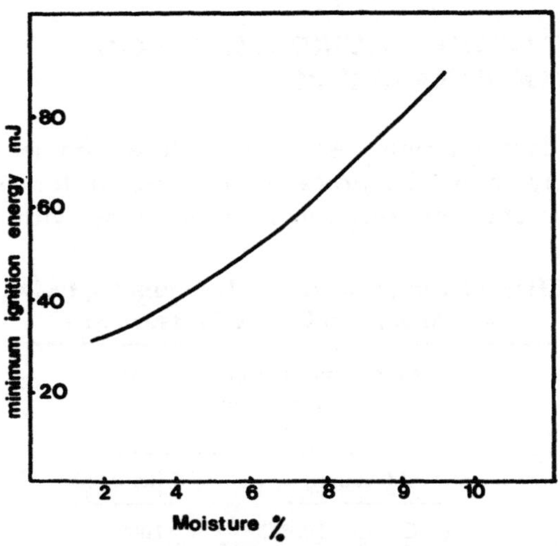

FIGURE 5.32. Effect of grain moisture on the minimum ignition energy.

TABLE 5.7. Effect of Dust Cloud Temperature on the Coil Temperature Required for Ignition.

Temperature of dust cloud (°C)	Coil temperature required for ignition (°C)			
	Lycopodium		Hydroxypropyl methyl cellulose	
	Undried	Dried	Undried	Dried
Ambient	920	930	950	950
50	910	920	940	920
100	890	900	860	870

The explosion parameters are also a function of turbulance and this may be one of the factors that causes the vertical-tube apparatus to give different results from a larger chamber. Turbulent mixtures tend to give higher explosion pressures than nonturbulent clouds.

It has also been found that there is a practically linear relationship between explosion data and the initial pressure at the time when the source was activated. This means that if an explosion occurs in one of two adjacent vessels the second vessel is probably pressurized ahead of the approaching flame front. The explosion in the second vessel is then likely to be more violent than in the first.

5.13. EFFECT OF HOT ENVIRONMENTS ON IGNITION PARAMETERS

The standard ignition tests are made at ambient temperature, however, many industrial processes operate at temperatures above ambient. Since chemical reaction rates increase with temperature it

TABLE 5.8. Effect of Temperature on Maximum Explosion Pressure Rise and Maximum Rate of Pressure Rise.

Substance	Maximum explosion pressure rise (kN/m²) P_{max}			Maximum rate of pressure rise (kN/m²) $(dP/dt)_{max}$	
	Measured		Calculated		
	20°C	100°C	100°C	20°C	100°C
Terephthalic acid	653	506	491	37.9	53.1
Giberilic acid	623	496	468	66.2	77.2
Lycopodium powder	707	534	533	59.3	69.6

TABLE 5.9. Effect of Temperature on Minimum Ignition Energy.

	Minimum ignition energy (mJ)	
Temperature °C	Lycopodium	Hydroxyproplyl methyl cellulose
Ambient	90	160
50	50	90
90	20	40
180	12.5	—

might be expected that explosion hazard will increase as the temperature rises.

Gibson and Rogers (1980) have tested this using a furnance apparatus modified so that the dust and air entering the tube were preheated to the tube temperature. A hot-coil ignition source and an electrostatic spark were tested; pressure tests were made in a modified Hartman bomb.

The results of the three tests are shown in Table 5.7–5.9. It can be seen that the coil ignition temperature decreased slightly as the ambient temperature increased, and that the decrease was similar for dried and undried samples. The ignition energy decreased markedly for the two dusts tested and the rate of pressure rise also increased significantly. The maximum pressure rise in the bomb decreased with increasing temperature.

So far the tests have not been carried out on an extensive range of powders, but it certainly cannot be assumed that ignition properties of powders at elevated temperatures are the same as at ambient temperatures.

5.14. SUMMARY

An extensive series of tests have been built up to measure parameters associated with the explosibility of powders and dusts. Sources of data are listed in Appendix A3. None of the published figures can be considered to be hard and fast constants for the material concerned, because there may be variations of 100% or more between different samples of the same material if particle size, agglomeration, or moisture content change.

Different test methods can also lead to very different measures of most parameters, and factory conditions such as temperature and humidity can strongly affect the dust explosibility. The published data of minimum energy and temperatures refer only to the particular test conditions used and the results of an explosion can also be considerably affected by test-apparatus geometry.

The use of published data for design of plant therefore requires careful interpretation. Often it is required only to classify the material into a broad group (such as when designing venting requirements) and published data may then be sufficient. However, in some cases, such as in the design of driers where maximum temperatures may be close to published ignition temperatures, individual tests on the particular powders concerned may be required.

6

Design of A Processing Plant for Safety

6.1 FORMAL SAFETY ANALYSIS

In recent years a number of large explosions and serious near-misses have occurred which have alerted management to the importance of design to minimize hazard and loss. As a plant becomes larger and more complex the task of ensuring that all eventualities are foreseen becomes more difficult. The result has been the development of a formalized approach to hazard analysis and loss prevention which is fast becoming a new science in its own right with a new vocabulary and mathematics.

In the last analysis, anticipation of the possible faults that can lead to a hazard relies on the imagination of the designers. The value of formal hazard studies is to provide a systematic stimulation to the imagination so that no potential hazard is overlooked. Certainly some form of detailed analysis is extremely important. One major international chemical company estimates that 4% of the total design cost is spent on analysis of hazards.

A number of separate hazard studies can be identified which are necessary before bringing a new plant into production (Gibson, 1976). The first occurs during the exploratory studies and has the aim of gaining an overall view of the dangerous properties of all materials that will be handled in the process. This includes explosibility, toxicology, chemical reactivity (particularly compatability with air and water), corrosiveness, temperature effects, etc. Many of these properties will be available in the literature or in computerized information services but others will require laboratory testing to simulate the circumstances likely to prevail in the plant.

The effect of possible impurities or accidental addition of other materials handled should also be considered at this stage. The study of the basic properties of the materials and products is recommended at the very first stages of plant design because it may have implications on the route chosen for the process. When the route of the process is defined, a more detailed analysis is carried out on each process, following the basic flow-sheet design of the plant, to consider the hazards associated with each element. The preceding studies are normally carried out before final approval for the plant is granted.

Having obtained approval, the interactions of the separate processes to be installed are reviewed in order to design a safe layout and define the relative positions of different items of plant. This includes not only siting potentially hazardous processes away from each other and in a position where they can vent safely, but also ensuring that there are adequate escape routes with room for firefighting appliances and isolating offices and staff working areas from potentially hazardous zones.

When the plant layout and flow sheet are defined, an operability study can be carried out. A formalized technique for this has been developed by ICI (Lawley, 1974). Seven key words are applied to every pipeline vessel and process in the system, so that all possible deviations from the norm are considered. Following this exercise, all operating limits are defined together with the safeguards necessary to ensure they are not exceeded.

Considerations of the hazards of an operation does not end with the conclusion of the design stage. Once the plant is built it is necessary to ensure that the design aims have been achieved and that all the details of safety design, (fire extinguishers, grounding procedures, temperature checks, etc.) are correct. After some months of operation a further study is required to ensure that no changes have been made to improve operational efficiency which might have adversely affected safety. During operation it is necessary to ensure that all changes made to the process or to the products handled are analyzed in detail from a safety viewpoint. This procedure is time consuming but many accidents occur because a process or product is changed. For example, addition of an additive to improve powder flow may affect sensitivity to static or the temperature sensitivity of the product. Increasing process velocities may increase electrostatic charging and will certainly cause additional heating.

Process safety requires constant review because minor day-to-day changes may frequently be introduced without a full study. For example, a recent fire was caused because a customer required a material in plastic buckets rather than metal bins. This isolated the electrostatically charged product which sparked to the ground.

Finally, it is very important that a full analysis is carried out of all minor accidents and near-misses to ensure that the circumstances could not arise again and cause a serious accident.

6.1.1. Operability Studies

An operability study is a formalized review of the operating limits of a process in which all possible deviations from the norm that could be hazardous are forseen. A number of techniques have been developed but one of the more common is the KEY word system developed by Lawley of ICI (1974).

In this system seven keywords are applied to every pipeline, process, and vessel in the system. The keywords are NO/NOT, MORE, LESS, PART OF, AS WELL AS, OTHER THAN, and THE REVERSE OF. To illustrate the use of these keywords consider a mixing process in which a number of materials (powders and liquids) are poured into a solvent. First the consequences of NOT adding each material in turn is considered. If, for example, one material is an antioxidant or an inert stabilizer, omission could lead to a hazardous situation. The possible causes of the fault condition must be identified and the consequences of the fault, both to the operation itself and to the subsequent parts of the process, are then evaluated. The actions required to prevent the fault can then be decided and the study can move on to the next keyword.

The effect of MORE or LESS for each material is calculated and the words MORE and LESS must also be applied to the process itself; for example, more heat, more speed, more solvent. AS WELL AS denotes additional activities other than those intended, including in this example accidental addition of some other component, or dropping something into the mixer. REVERSE applies mostly to activities such as operating valves or chemical reactions for which a distinct reverse to the intention can be defined. OTHER THAN denotes complete substitution; for example, the addition of some material handled in the plant other than that intended.

Each major keyword can be given subheadings depending on the process. For example, temperature, pressure, volume flow, or level can be used as subheadings for MORE or LESS. Special instructions such as the order in which the additives are made are included, because each addition is treated as a separate element. If A, B, and C must be added in that order than the step of adding A to C is then studied as part of NOT adding B or of adding C OTHER THAN B. Table 6.1 shows the use of guide words in defining some typical problems. Table 6.2 gives a list of deviations from the norm that might have to be included.

It is usually recommended that this kind of study be carried out by a team of people with varying expertise who will consider each aspect of the process—mechanical, electrical, chemical, etc. The team is led by a study leader who takes each process through every logical step. An operability study is time consuming and therefore expensive but the vast

Table 6.1. Guide Words in Operability Studies.

Guide word	Type of deviation	Typical problems
NOT	No flow	Blockage; pump failure; valve closed or jammed; leak; valve open; suction vessel empty; delivery side overpressurized; vapor lock; control failure
REVERSE OF	Reverse flow	Pump failure; NRV failure or wrongly inserted; wrong routing; delivery overpressurized; back-siphoning; pump reversed
MORE	More flow	Reduced delivery head; surging; suction pressurized; controller saturated; valve stuck open; leak; incorrect instrument reading
	More temperature, pressure, viscosity	External fires; blockage, hot spots; loss of control; foaming; gas release; reaction; explosion; valve closed; loss of level in heater; sun
LESS	Lower flow	Pump failure; leak; scale in delivery; partial blockage; sediment; cavitation; poor suction head; process turndown
	Low temperature, pressure, viscosity	Heat loss; vaporization; ambient conditions; rain; imbalance of input and output; sealing; blocked vent
PART OF	Change in composition	High or low concentration of mixture; additional reactions in reactor or other location; feed change
AS WELL AS	Impurities or extra phase	Ingress of contaminants such as air, water, lube oils; corrosion products; presence of other process materials due to internal leakage; failure of isolation
OTHER	Activities other than normal operation	Startup and shutdown of plant; testing and inspection; sampling; maintenance; activating catalyst; removing blockage or scale; corrosion; process emergency; safety procedures activated; failure of power, fuel, steam, air, water, or inert gas; emissions and lack of compatibility with other emission and effluents.

majority of accidents that do occur might have been foreseen by this type of study.

6.1.2. Fault Tree Analysis

The use of fault trees provides another useful formal approach to hazard analysis. In a full process the fault tree analysis can become highly complex; only a basic introduction is given here.

Table 6.2. Types of Operating Deviation.[a]

Major operating limits:	Flow; temperature; pressure; level; chemical reactivity; mechanical stresses
Other operating limits:	Corrosion; erosion; resistance; fouling; cavitation; vibration; hammer; loadings; expansion; contraction; thermal/mechanical shock; cycles of activities; environmental factors
Material physical characteristics:	Viscosity; miscibility; melting/boiling points; density; vapor density; phase; appearance; particle size
Chemical composition:	Hazard characteristics of mixtures
Reactions:	Extent and type; side reactions; catalyst behavior (e.g. activity, toxicity, reaction, regeneration, decomposition); planned/unplanned reactions; contaminants including materials of construction; corrosion products; catalyst; reactor runaway; explosion; combustion
Time:	Contact time; sequence; design cycle
Local effects:	Distribution; mixing; hot spots; overheating; resonance; stress on bearings/shafts; lubricating faults; vortex generation; blockage; slugs; sedimentation; stagnation; adhesion; crushing; grinding; separation; surging; siphoning
Failure to contain materials:	Spillage; leakage; vented material
Construction:	Defective materials of construction; plant incomplete; plant unsupported; plant not aligned; plant not tight; plant not level; plant not clean
Major deviations:	Start up/shutdown; maintenance/inspection; planned changes in normal operations; supply/equipment failure; demand change; unplanned ignition source; control disturbance; loss of communications; human error; climatic effects

[a] Where appropriate, both static and dynamic effects must be considered.

A simple fault tree which looks at an explosion caused by an electrostatic spark during a mixing operation is shown in Figure 6.1. The process is illustrated in Figure 6.2. Under correct operation the liquid is poured manually from a metal drum to which a ground clip is fitted. The operator stands on a conducting floor wearing antistatic footwear and there is an extraction system which prevents a hazardous concentration of vapor outside the mixing vessel. The failure to be analyzed is an explosion. This is placed at the top of the tree and is known as the *top event*.

If an explosion occurs due to an electrostatic spark there must be both a flammable mixture and an electrostatic source present. These two events must simultaneously occur and they are shown on the tree joined by the symbol  to indicate a logic AND. A flammable vapor may occur

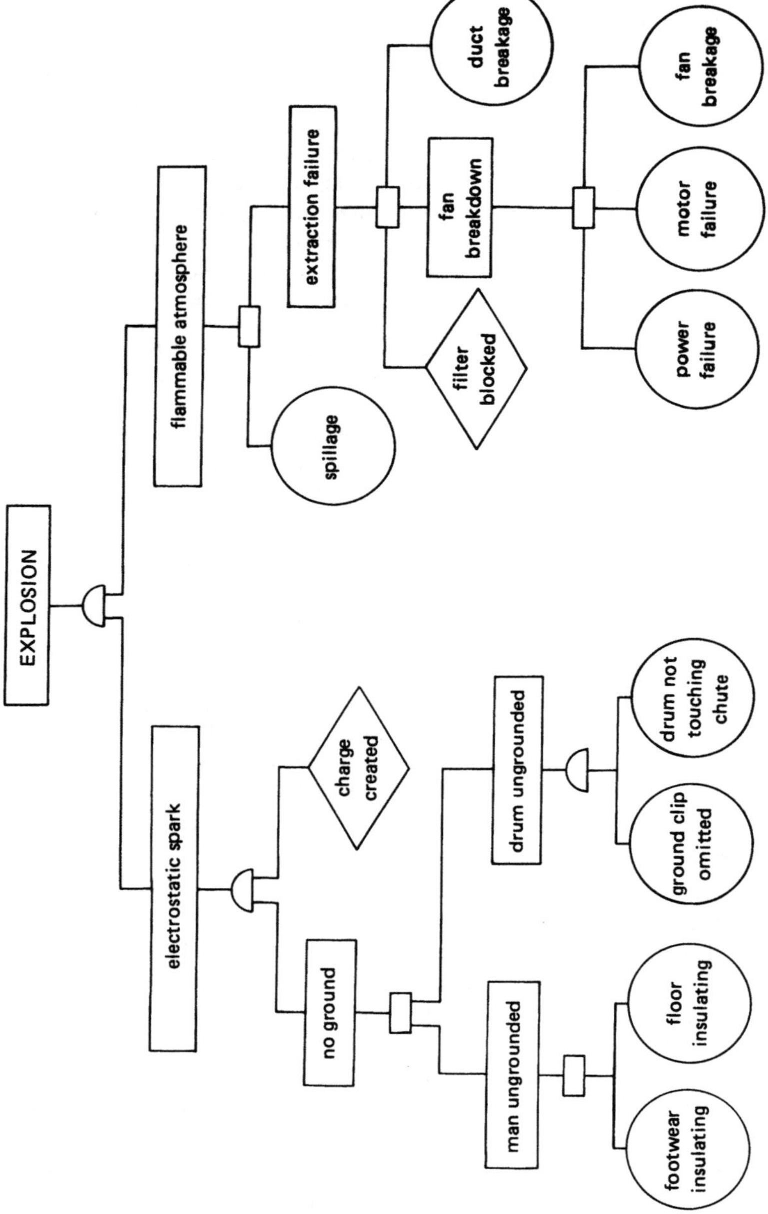

FIGURE 6.1. Fault tree analysis of a mixing operation.

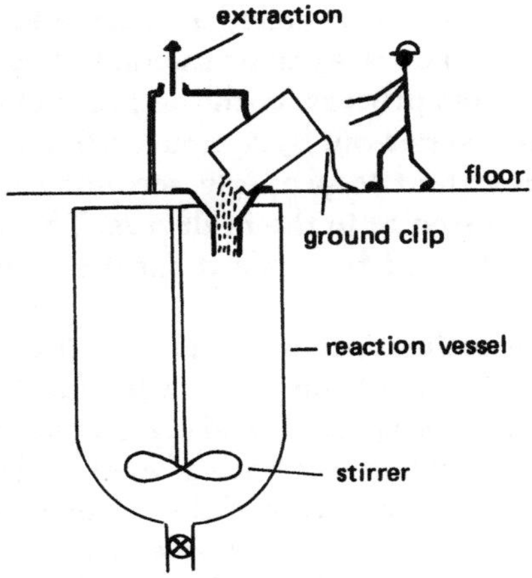

FIGURE 6.2. Schematic diagram of a mixing operation.

either if there is a spillage or if the extraction system fails. These two events are both independently capable of causing a vapor cloud and so are joined by the symbol ▭ to indicate an OR logic gate. The failure of the extraction system may have been caused by the fan stopping, the filter blocking, or the duct breaking, and so on.

Symbols are also allocated to indicate whether an event is to receive further analysis. The event is surrounded by a rectangle when it is further developed and a circle when the meaning is clear and no further development is required. In this case it is known as a *base event*. When an event requires further development, but this is not carried out within the tree, it is surrounded by a diamond.

Having drawn up the basic tree, a probability must be allocated to each fault. For example, if the extractor fan is known to break down twice a year and 400 mixing operations are carried out each year the probability is 1 in 200 operations.

Quantifying probabilities for operations and faults in which human error is involved is more difficult, but with experience probabilities can be allocated (Kletz, 1980). For example, observation may show that the operator fails to apply the ground clip 1 time in 10, and the safety relies on him resting the drum on the grounded chute. The number of times that this occurs can also be observed.

The failure rates of machinery or equipment, such as the antistatic footwear, is obviously a function of the frequency and effectiveness of routine maintanance. This too must be included.

When the tree is complete it can be analyzed either to determine the frequency of the top event or the alternative causes. Each combination that can cause the top event is called a *cut set*, and the frequency with which each combination occurs can be calculated by the laws of Boolean algebra. In a complicated process, both the production of the fault trees and their analysis are carried out by computer. (Fussel and Rovers, 1974). It is particularly important to properly account for failure events that occur at more than one point in the fault tree. These *common mode failures* are usually failures derived from plant services such as power, air, or steam.

Some degree of estimation and approximation is inevitable in the allocation of probabilities, but the value of the fault tree is that it clearly identifies the weak links of a process and gives a quantitative indication of the likelihood of a dangerous sequence of events. This can be combined with an analysis of the likely effects of the top event in order to evaluate whether the degree of risk is acceptable. More details of fault tree analysis are given by Lees (1980) and Andow (1980).

6.1.3 Event Tree Analysis

Event tree analysis is similar in concept to fault tree analysis but the top of the tree is a single event. The tree is used to evaluate the alternative ways in which the whole system can fail as the result of one single failure event. An example taken from the work of Andow (1981) is

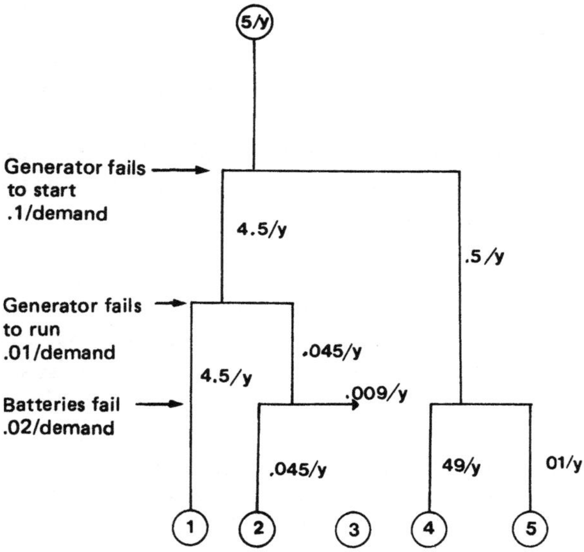

FIGURE 6.3. Event tree analysis.

shown in Figure 6.3. The initial failure in this example is the loss of power from a substation. When power failure occurs a diesel generator should automatically start using control power from a battery system. If the diesel does not start or does not run long enough, then the batteries provide sufficient power for safety.

By convention the event tree is drawn in the way illustrated in Figure 6.3. The failures are listed on the left-hand side of the page and the failure-occurrence branch goes to the right. The figures by each branch are the frequency of that occurrence per year or per demand. The total of all path frequencies at the bottom of the diagram must equal the frequency of the initial event. Sometimes the frequency of a path becomes so low that it is pointless to consider any further failures. This is shown by path 3. Termination is indicated by an arrow.

Event tree analysis is not as widely used as fault tree analysis but it has useful application to systems where there are multiple redundant means of satisfying a requirement.

6.1.4. Failure Mode and Effect Analysis

This type of analysis is normally carried out in order to study plant reliability. However, as system safety obviously affects reliability the study is also a type of formal hazard analysis. In failure mode and effect analysis every possible failure and its effect on the system as a whole is considered. Table 6.3 gives the headings under which the analysis can be carried out together with an example applicable to the sanding of a plastic sheet.

Failure can apply not only to equipment but also to human beings. Human failure modes are (a.) failure to perform a task; (b.) incorrect performance of a task; (c.) introduction of an additional step that should not occur; (d.) performance out of sequence; and (e.) failure to complete the task in the allotted period.

TABLE 6.3. Failure Mode and Effect Analysis.

Component under analysis:	Pump for cooling water to sanding drum
Error mode:	Stopped
Effect on other components:	None; sanding continues
Effect on whole system:	Sanding drum overheats
Hazard probability:	Reasonably probable
Method of failure detection:	Temperature probe on drum
Contingency provision:	Machine stopped if temperature high
Hazard classification:	Severe hot particles are created that could cause fire

6.1.5. Accident Investigations

If an accident or a near-miss does occur the sequence of events which combined to cause the incident are analyzed in detail. The aim is both to investigate how the initial hazard analysis failed and to ensure that the circumstances are not repeated.

A detailed inspection of the results of an incident is very important from two points of view. First analysis of the results of a explosion can often give considerable information about the causes of the accident. For example, clean patches under blackened ground clips indicate that they were in fact in position at the time of the ignition. The extent of carbon deposits and the magnitude and direction of the explosive blast also help identify the cause and location of the ignition source.

Analysis of the results of an explosion is necessary in order to add to the general store of information, as a guide to future policy on plant location and design, and as an aid to hazard analysis. Four aspects of the results of an ignition can be analyzed. The extent of flame spread, thermal radiation effects, blast effects, and missile projection.

The type of damage caused by a flame depends on the duration of the flame; therefore analysis of the damage gives information on the manner in which the flame spread. Radiation damage may be distinguished from direct flame damage because it tends to affect mostly the side of the object facing the heat source. Shadow effects may also be observed. Radiation effects depend on the temperature and duration of the exposure. Data are available on the response of various materials to radiation (Lawson and Simms, 1952; Stoll and Chianta, 1971).

A blast wave from an explosion will create a direct pressure-shock wave and also an airflow that gives high-velocity winds. Damage to vessels, walls, and windows can give estimates of overpressure which can then be plotted as a function of distance from the explosion source. Damage also indicates directional effects.

The position of missiles projected by the explosion can also give very useful information. If the mass, approximate shape, and initial location of the item is known, an estimate can be made of the forces to which the objects were exposed during the explosion. The information gained from analysis of all the effects will enable the probable location and force of the explosion to be identified.

6.2. HAZARDS ASSOCIATED WITH INDUSTRIAL PROCESSES

In order to analyze the safety of an industrial process each individual operation involved must be considered separately. The ignition sources that may arise in each operation must be identified and the methods of protecting against ignition defined. Although the operations are considered individually the ignition sources which may be carried from one plant site to another must not be forgotten.

Operational faults in one part of a plant may cause a fire hazard in another. For example, a buildup of powder due to failure of a metering device could overload a conveyor belt, causing it to overheat. Hot particles may be created in a dryer and carried through the system to a storage vessel where they cause an ignition.

In the following sections the problems specific to particular processes are described but the special problems that arise because of the juxtaposition of the different elements must be added to the analysis in each situation.

6.2.1. Drying

The drying of powdered products is particularly prone to fire and explosion hazards because it is a process in which an external heat source is applied. The most common cause of ignition tends to be hot particles. These may be caused by recirculating air containing residual dust particles, through the heater units or by hot surfaces that initiate thermal decomposition. If the process involves drying a solvent it is particularly hazardous because of low ignition energy and high explosibility of hybrid mixtures of dusts and flammable vapors.

There are a number of different types of drier commonly used in industry each of which has its own particular problems, however, many of the general rules are applicable to all driers. In a drying process it is particularly important to accept published ignition data only with care. The ignition properties of dusts, particularly those concerning the ignitability, are measured under specific conditions that may not apply in a drier. For example, the miniumum ignition energy is measured at room temperature and a dust may be more easily ignited when warm. Minimum ignition temperatures are often measured under isothermal conditions and adiabatic measurements, such as produced by accelerated rate calorimetry, have been shown to give ignition temperatures which may be 50–60° C lower than found with isothermal measurements made by differential thermal analysis. This may be a critical difference of around 30%.

It is also very important to ensure that the dust tested is in the same chemical state as that in the drier. For example, dust layers in a drier may contain deposits of volatile materials that increase the layer ignitability. The state of oxidation of the dust may also differ once the material has been exposed to the atmosphere while awaiting tests.

Self-heating is one of the more common causes of ignitions in drying systems and it is important that the thermal properties of the material under the conditions which prevail in the drier are known. The temperature of all surfaces in the drier at all stages in the process should be less than the temperature at which self-heating begins. But if this is not possible the temperature of the layer, the thickness of the deposit, and the time for which it remains in a heated condition must be analyzed to ensure that the conditions are below the limits where self-heating can proceed to ignition. Identification of the conditions that can produce ignition is rather more difficult than finding the start of self-heating because it depends on achieving the optimum conditions for the reaction to proceed. In general, driers should be designed so that there are no ledges or crevices where dust can accumulate for long periods giving slow self-heating an opportunity to proceed to an unsafe limit.

Particles of powder may also overheat if they are able to penetrate the filter to the heating system. They may then be emitted with the drying air having been directly heated. Deposits of dusts in areas which may get hotter than the main drier are a particular hazard. Spray driers may throw particles onto the top of the chamber where the hot air enters and heated layers falling from this position can ignite a dust cloud in the drier.

Electrostatic discharges will be a hazard in a dust cloud if there is an isolated conductor on which charge can build up. A common example occurs in fluid bed driers, where the bag may be made of an insulating material clamped in place by a ring which is not directly grounded.

Some of the more sensitive dusts can be ignited by the spark from broken steel fibers which are incorporated into a fluid bed drier bag with the intention of reducing static. Although these are effective when new, they can present a hazard once the bag has been washed and the continuity of the wires is broken. Bags made of an insulating material are unlikely to present a direct hazard for dusts whose ignition energy is above 25 mJ but should be avoided with highly sensitive dusts and if flammable vapors are being removed.

Electrical equipment should be in an area which is not dusty, but equipment is sometimes necessary in the vicinity of a drier where it can become dusty. The equipment should then be designed to be sparkproof and dust tight. It is also important that dust settling on the outside of electrical equipment should not prevent it cooling properly such that the equipment and dust layer can overheat.

Inspection lamps should be properly designed for the purpose and, as with all lighting, the bulb should be protected by a dust-tight casing. The possible additional heating effects of the lighting should be included in consideration of the thermal stability in the drier.

Frictional heating is often cited as a cause of drier ignitions. This may take the form of general overheating. For example, if the material is not properly dried it may clog and overload an output screw conveyor. Impact of tramp metal can also cause ignition. Although the ignitability of a single impact is not proven, tramp metal can certainly cause problems when it scrapes along the plant or is broken up in a grinder or a screw feeder. Metal parts thrown off at high speed, such as from the atomizer in a spray drier, can create considerable heat by friction. The thermite reaction in which an impact involves rust and alloys containing aluminum, magnesium, and titanium can certainly ignite a dust cloud, and it has been shown in the laboratory that many common organic powders are sensitive to this ignition source (Gibson, 1968).

In all types of drier the same procedures may be followed to identify hazardous areas. First, the regions where a cloud of dust may exist, either during normal operation or if there is a fault, are defined. Since it is never possible to ensure that all ignition sources are eliminated, it is necessary to protect all parts of plant where a dust cloud may exist and ensure that ignition in these regions cannot spread throughout the plant.

Protection may take the form of containment, inertion, venting, or suppression. For example, large spray driers and fluid bed driers are most often vented but smaller spray driers may contain suppression systems. Flash driers (pneumatic conveying driers), in which the material is dried by dispersing it into a gas stream and passing it through ducting, may be protected by a combination of vents (where these are practical) and a strong structure to withstand explosion in regions (e.g., a long vertical duct) where effective venting is impractical. Suppression systems may also be used in this type of plant.

Having considered the areas where a dust cloud may exist, the hazard of dust layers is studied next. Areas where dust may be able to accumulate in both normal circumstances and under fault conditions must be taken into account. This includes the walls and base of a spray drier, bends in a pneumatic drier, and horizontal regions of a pneumatic drier if the airflow rate decreases. Consideration must be given to the temperature which might be expected in the areas where the dust can accumulate, again both under normal and fault conditions.

The types of fault that might produce a dust cloud in a region normally free from a cloud or that might cause temperature above the normal operating condition should be considered. In some driers it may be possible to accumulate flammabe vapors from drying solvents or from decomposition of heated powders. Very great care should be taken in these circumstances. Many driers have a recirculating air system in which air is filtered by cyclone scrubber or bag filter and the fine dust separated out is fed back into the main system. Although the dust-recovery section is likely to be cooler than the main drier, it tends to contain the driest and finest dust and hence the most explosive material. Precautions must be taken to ensure that dust does not settle out in ducting and the driving fan must be downstream of the final dust-collection point. The possibility of hot particles from the drier being carried through into the dust-recovery system cannot be excluded; therefore some form of protection should be applied to the recovery system. Dust-collection systems are discussed in more detail later in this chapter.

Case Histories—Fires and Explosions in Driers

Grain Dust Explosion in Maltsters, 1976

A fire of unknown cause started in a grain drier and initiated a series of grain dust explosions culminating in a large single explosion. The total damage was over £400,000 and two men were involved.

The men were originally involved in maintenance work and were walking back to their workshop when they noticed a smell of burning coming from a drier. While they were attempting to extinguish the fire, a large explosion occurred. Considerable damage occurred throughout the buildings. Flying debris cut overhead cables and electricity and gas supplies were therefore cut off. However, power was later restored to selected areas to allow circulating fans to operate and prevent other driers from overheating.

Analysis of the explosion pattern showed that the blast travelled via underfloor ducting and up staircases and the passenger lift. Dust had been allowed to accumulate to over 2 cm deep in places on floors, ledges, and pipework throughout the building.

Smouldering persisted in these deposits some time after the main fire had been extinguished. Although the source of ignition was not identified, beyond doubt, damage was consistant with a fire smouldering in the grain.

Decomposition During Drying of Diphenyl Black (Dyestuff)

The dyestuff was dried at 100°–105°C in a circulating air drier. The air fan and the steam were shut off at 4:00 pm. The dyestuff remained in the drier overnight. The temperature chart showed a temperature rise starting at 8:00 pm. The decomposition of the dyestuff was first discovered on the following morning when the oven was opened.

Fortunately no one was injured. However, the cause of the incident is unknown. According to tests made, the dyestuff belongs to Hazard Class Tr 0, maximum allowable wall temperature 140°C.

Explosive Decomposition While Drying

300 g of the monoazo derivative of a nitro-amino-phenol and a pyrazolone were being dried under vacuum at 110°C when a violent detonation occurred. The drying oven was heavily damaged. No one was injured.

Small quantities had previously been dried at 110°C without evidence of decomposition. A possible explanation of the difference in behavior is that the harmless batches filtered at pH 7.5 and the explosive batch at pH 8.8. In addition, a check revealed that the surface temperature of the trays in the drying oven differed from the recorded temperature by 20–30°C. Such temperature differences can suffice for a spontaneous decompostion in the case of critical products.

Explosion in a Poultry Feed Drier

An explosion with a force of 200–300 lb of TNT occurred in 1976 causing £1 million damage. A poultry feed additive (3:5-dinitro-*o*-toluamide) was dried in a batch vacuum dryer with steam supplied to the heating jacket at 150°C. Drying was completed after 24 h at which time the steam was turned off and the vacuum broken. This was Saturday afternoon. The batch was left in the dryer to await the Monday morning shift. On Sunday afternoon the dryer exploded killing one man.

The Health and Safety Executive report noted that neither previous experience nor available thermal data had suggested that the compound was thermally unstable at the drying temperature of 120–130°C. Examination of the compound by differential thermal analysis had shown it to be thermally unstable at higher temperatures, but it was not known that the compound could self-heat to destruction if held under adiabatic conditions of 120–130°C for 24 h. This was subsequently shown by thermal stability tests using accelerated rate calorimetry carried out under conditions similar to those prevailing in the drier.

Sugar Refinery: Fire Caused by Dust Explosion 1977

A fire at the premises, causing £100,000 estimated damage, was the result of a hot particle in the sugar beet pulp drying equipment igniting recirculated dust in the outfall housing. This resulted in an explosion and subsequent ignition of the factory wall cladding, the dust on the roof beams, and of the pulp (about 4 tonnes) in the process of being dried.

A modified system of dust extraction, which included the recirculation of dust extracted from the outfall housing, had just been installed prior to the start of the sugar beet season. The equipment had only been running for 4 h, following the modifications, when the explosion occurred.

An investigation carried out by the Health and Safety Executive, Fire Brigade, and the company revealed that due to the modification, fine dust was being introduced into a brick enclosure situated at the point where the tunnel drier joined the conveyor system. The enclosure was not equipped to cope with this and the explosion occurred.

The outfall housing was a large brick-built chamber where the dried pulp was conveyed from the drying machinery through ducting to a different part of the factor.

The explosion severely damaged the brickwork of the outfall chamber and injured a workman who had to be taken to the hospital but was not detained.

The "housing" was situated inside a steel-framed, single-storied factory (40×40 m^2 approximately), built of plastic and bitumen-covered steel sheeting (the roof had an inner lining of asbestos insulating board) and a concrete floor.

Fire and explosion severely damaged the outfall housing structure. Fire, heat, and water damaged, in varying degrees, the external cladding of the factory and its roof. Drying machinery and ancillary equipment were also damaged.

Fire Caused by Hay Drying Unit on a Farm 1977

A spark from a heater unit, used for drying hay, ignited dust or other combustible material which in turn ignited the timber in the main channel of a detached, single-storied shed approximately 45 × 30 m^2. From the main channel the fire spread rapidly, destroying about 200 bales of hay, 400 tonnes of barley and wheat grain, and damaging, by fire and smoke, 1200 tonnes of barley and wheat. A large proportion of the roof (double sheeting of asbestos cement, insulated with glass fiber) was damaged by fire but the brick walls and the floor were unaffected. The estimated loss was £110,000.

The heater unit worked in conjunction with a diesel-driven extraction unit. Both units were situated outside the shed, blowing and drawing air through 1-m diameter holes in the gable end walls. These units had been operating continuously for about 3 weeks (a routine daily check was carried out) before the fire started. The fire was discovered by the farm bailiff who had checked the units at midnight and had noticed nothing wrong.

The dry timber ducting and hay rapidly spread the fire which was fanned by the air from the moisture-extraction unit. This was still operating when the Fire Brigade arrived about 20 minutes after being alerted. The fire's intensity visibly decreased when the unit was shut off.

Following the incident it was recommended by the Fire Brigade that (a) the moisture-extraction unit should be fitted with a small gauge-protection grill when used with a heater unit, and (b) when the heater unit was used for hay drying, a series of baffles should be fitted to prevent any hot particles from being emitted as a result of spalling on the inside of the heater.

6.2.2. Grinding

A major hazard in grinding machines is the presence of tramp metal or stones which produce heat and sparks igniting the material being ground. Magnetic separation can be used to remove metallic tramp but this can not remove absolutely all foreign bodies. The problems of foreign bodies in grinding machinery, although severe, are not the only cause of ignition. The grinding process may create considerable quantities of heat and desposits of heated dust may ignite if they are allowed to build up.

Micronizing and grinding also produce a high level of static charge. The danger of static is once again not in its formation but in its accumulation, particularly on isolated conductors. For example, the practice of mounting metallic sieves following a grinding operation on antivibration mountings may create an isolated conductor. Powder transferred directly from a grinding machine into bins or boxes may create a thermal hazard if the temperature is not well below the quoted ignition temperature, or it may create an electrostatic hazard if a conducting powder is poured into polyethylene-lined bins.

It is often difficult to effectively vent or suppress a grinding machine because of the complicated structure and moving parts. Normally the installations are small enough for it to be feasible to build the structure to withstand an explosion but with very sensitive dusts inerting may be the best method of protection.

6.2.3. Transport and Storage

6.2.3.1. Belt and Bucket Conveying

An analysis by Factory Mutual Insurance has shown that the three most common causes of fires on conveyor systems are frictional heating, welding and maintenance accidents, and the dumping of the overheated material on the belt.

Overheating of the belt by friction may occur when the belt loses its traction and slips, or because it is loose or misaligned. Bearings in conveyor systems may overheat due to age, lack of maintenance, or the ingress of dust. This type of problem is most severe in the very large plants, similar to those found in the grain industry, where there may be a large number of belt systems and few men to notice the danger signals of overheating.

Modern plants are built with overload protection on motors and instruments to measure belt speed and detect pulley slippage. Conveyor belt systems tend to be open and so housekeeping is commonly poor and layers of dust accumulate around the belt and on the surfaces of motors and machinery. The constant presence of the powder leads to an attitude of acceptance in the workforce, so less care is taken to clean up effectively for maintenance.

In most of a belt conveying system the powder itself does not move with respect to the surface on which it lies and electrostatic charging therefore tends to be low. However, high-speed insulating conveyor belts can produce considerable static, particularly on isolated metal splices. Other isolated conductors in the system which may accumulate charge are metal buckets on insulating belts. Insulating conveyor belts are nowadays rare but some old equipment still operates with the old style of belting.

Metal buckets can also create a hazard when they become detached from the belt and fall down the elevator creating sparks. Plastic buckets are better if the belt material is insulating and also prevent impact sparks. The ideal material would be an antistatic plastic since this avoids both the danger of impact sparks and the possible hazard of a conducting powder in the bucket forming an isolated conductor. However, the latter is not likely to be a serious problem if the belt and bucket surface are coated with the conducting dust which prevents charge from being stored easily.

Case History

Sugar Refinery 1972

In a refinery, sugar was conveyed through all storeys of the 13-storey building by two elevators. About 1½–2 min after one of the elevators had been switched on following a 9-day shutdown, an explosion occurred. It appears that sugar had accumulated in the elevator pit from another part of the plant via a flap valve. This put abnormal strain on the elevator which had to dredge through this accumulation causing a misadjusted tensioning device to fail. The elevator buckets now running severely out of alignment rubbed hard against the bucket guides and were struck heavily by the sprocket wheels. The heat generated by friction between the metal surfaces was sufficient to ignite the cloud of sugar dust. The estimated total loss was £70,000.

The elevator duct and parts of the associated ventilation and sugar sieving plant were severely damaged in the explosion. Some window frames and many windows were blown out and internal partitions were damaged by the blast. Fire damaged 3500 sacks and a similar number were damaged by water. Two employees were severely burned.

6.2.3.2. Pneumatic Transport

In a pneumatic transport system powder is dropped into an air stream and blown through pipework or pulled through by suction. The density of the dust/air mixture varies between a minimum of 0.2 kg/m^3 for dilute-phase conveying and 1000 kg/m^3 for dense-phase conveying. At the start and finish of transport even dense-phase systems may have clouds within the explosive limits. In pneumatic transport the powder rubs against the walls of the conveying tubes with a considerable velocity; thus there tends to be a high static charge created. Provided all sections of the pipe are conducting and are grounded there can be no charge accumulation in the pipework itself, but when the powder is collected after conveying very high fields can arise if the powder is not sufficiently conducting for the charge to leak quickly. Although ignitions due to electrostatics occur more commonly at the end of the transport process rather than in the pipes, great care should be taken that flexible sections inserted to give mechanical isolation of load cells, or to assist transport, do not create any electrical isolation leading to ungrounded conductors.

Pneumatic systems should be designed so that dust does not settle out during transport. This is achieved by high velocities and gradual bends. Dead legs should be avoided so that dust cannot accumulate and self-heat. Some powders may coat the walls of tubes either by impact adhesion or by electrostatic forces. The possibility of these deposits heating (for example, if a hot product is conveyed) must be considered.

Heated particles created during grinding or drying may be carried in a pneumatic transport system and fanned to a glow by the high air velocity. These can then cause an ignition in the storage or collection system at the end of the pneumatic transport. Tramp metal in pneumatic systems may cause frictional heating or sparks as it is passed through the system and should be eliminated as far as possible.

6.2.4. Dust Extraction and Collection

Dust extraction is fitted to many plants to prevent an explosive density of flammable dust. There is a tendancy to assume that because the dust-extraction system prevents dust buildup in the plant the explosion problem is eliminated. However, design considerations must take account of the possibility that the extraction system will break down as well as considering what will happen to the dust when it is collected.

The minimum ignition concentration of most flammable dusts is ~ 5–$25\ g/m^3$. The air flow in the extraction system should be sufficiently high so that even in the worst case the dust density in the ducts is below the minimum ignition concentration. It is also necessary to ensure that the dust does not settle out in the duct, and this in itself requires a velocity—depending on pipe diameter—of 5–25 m/s. If the pipe diameter is too large this may be difficult to achieve. Sharp bends in extraction ducting should also be avoided because this tends to lead to a dead space where powder drops out. Horizontal sections should be as short as possible for the same reason. In some processes the extraction system may also collect volatile materials that form a deposit on the duct to which dust can adhere. If this happens frequent cleaning is necessary. Although the dust cloud in the duct itself may be below explosive concentrations, dust layers that self-heat may cause hot particles to be transported through the ducts to the dust-recovery system where there is frequently an explosive concentration of material.

The fault conditions of the plant to which the extraction system is fitted should also be considered. For example, if extraction is from a drier that is not drying efficiently the wet particles may build up on the inside of the duct, then later dry or self-heat.

If the duct necessarily must carry dust above the explosive density, suppression or venting should be fitted, or it should be designed to withstand and contain an explosion as discussed in Chapter 7. Inspection covers should be provided at intervals to allow inspection and cleaning of any deposits.

The dust-collection system may be one of a number of types. The most common are cyclones, bag filters, electrostatic precipitators, and wet scrubbers. In all dust-collection systems there is the possibility that the dust can be disturbed to form a cloud. The particles involved in an extraction system tend to be small and hence more explosive than the larger sizes dealt with in the main process.

A fault resulting in a fire in almost any other part of the plant leads to the risk that burning particles will be collected by the dust-extraction system and carried through the ducts to the collection area. Dust-collection systems handling a flammable material should therefore always be considered as a potential explosion hazard. They should be sited in a safe place so that an ignition does not spread to other plant sites and fitted with suitable venting or suppression. The practice of siting banks of dust collecting systems close to each other and venting towards each other is *not* recommended.

6.2.4.1. *Cyclones*

The outlet air duct of a cyclone may be regarded as an explosion vent only if it is of suitable area and not followed by ducting or a bag filter which provides a resistance to the escape of the explosion. In this case a vent can be fitted in the roof of a cyclone near the inlet duct or the cyclone should be made sufficiently strong to withstand the explosion. If the cyclone is strong, the explosion pressure will travel down the inlet and outlet ducts, both of which are required to be strong. Cyclones should be fitted with a rotary valve at the base to act as an explosion seal.

6.2.4.2. *Wet Scrubbers*

Wet scrubbers do not constitute a safety hazard since the dust cloud is damped down. However, the collected sludge when dried out could be a hazard.

6.2.4.3. *Electrostatic Precipitators*

In an electrostatic precipitator dust is intentionally charged by ions from a corona discharge originating from a high voltage applied to a sharp electrode. The charged particles are then collected by means of an electric field applied between flat plates. The chief danger is the possibility of arcing between the electrodes. For example, when the level of dust accumulation reduces the electrode gap a spark may occur.

Precipitation is seldom used for flammable dusts because of the difficulties of preventing arcing. The current from the power supply to the precipitator plates can be limited by a suitable electronic design which prevents sparks drawn from this source, but a discharge can occur whose energy W is given by

$$W = \frac{1}{2}CV^2$$

where C is the capacitance of the collector plates and V the voltage applied between them.

An explosive dust cloud may exist in the precipitator when the plates are shaken to remove the deposit or if part of the collected dust should be disturbed. Therefore, a precipitator is not used on materials where ignition energy is below the available spark energy in arc-over unless the atmosphere is inerted.

6.2.4.4. Bag Filters

Bag filters are the most common form of dust-collection system and frequently follow cyclones when the fine-dust load is high. Fabric filters contain a dust cloud during cleaning which is often automatically induced by vibration or reverse air jet. During cleaning an explosive concentration exists in the filter. Smouldering particles collected by the dust-extraction system, which lie on the fabric with air passing through them, are in a good position to continue to burn and spread fire to other dust until disturbance during cleaning creates an explosion.

Dust arriving at the fabric filter is likely to carry a high electrostatic charge. Manufacturers of bag filters have attempted to reduce this by means of metal wires woven into the fabric, similar to those in fluid bed driers. This enhances the electrostatic hazard when the wires break. Once again it is important that there should be no ungrounded metal within the filter, and bags made of a conducting material should be used with very static-sensitive dusts or in systems where flammable vapors may be extracted or may form on decomposition. There are a number of bag filter types that have antistatic properties which do not wash out and do not result in isolated conductors. For example, epitropic fibers have been developed in which carbon is incorporated into every fiber. There are also felt construction bags in which there is a mass of very-fine, short fibers made of steel or carbon which cannot become isolated from each other with age.

Vents on dust filters are frequently fitted on the clean side of the dust bag because it is here that there is normally a clear surface at the top of the filter of sufficient area for a vent. However, the dust cloud will occur on the dirty side of the bag and the bag itself may provide a considerable resistance to the escape of the explosion. The rupture pressure of the bag may considerably exceed recommended opening pressures for vents and hence lead to a much higher overpressure in the filter because of the rapid rate of pressure rise which occurs in a dust explosion. A vent should therefore be situated on the dirty side of the bag filter although this may need to be supplemented by an additional vent on the clean side. If vents are only provided on the clean side, the filter container should be built strongly to contain the additional pressure rise which occurs during the time taken for the bag to rupture. Allowance must be made for the vent to become partially blocked by the ruptured bag.

Air from dust bag filters may be recirculated back into the system to prevent heat loss. (This particularly applies to drying systems.) In this case, the possibility of dust escaping through the bag or bag damage resulting in dust passing through it must be considered. The recirculating system must be desired with the possibility of dust being present in the air stream kept in mind.

Case Histories—Extraction Systems

Metal Powders Plant, U.S. 1978

This plant used aluminum ingots to produce aluminum particles of various sizes: pellets, flakes, and dust. The ingots were atomized and then transported by an air conveyor system to primary and secondary cyclones for extraction of the various particle sizes. The resulting particles were deposited in bins at the cyclone bases.

Extensive precautions to limit damage in the event of explosion had been taken in the design of this facility. The ducting along which the atomized metal was transported had 18 × 4-in^2 explosion vents located approximately 10 ft apart along its entire length. The first primary cyclone separated out the largest particles, then the exhaust entered a secondary primary cyclone, and finally the smallest particles were fed into a "bag house." The cyclones had explosion vents, and the bag house was surrounded on three sides by a barrier wall designed to deflect explosive force away from occupied areas. The barrier wall was constructed of concrete blocks reinforced with concrete and steel. The bag house acted as a "seal" to limit the number of particles emitted to the atmosphere.

As an additional precaution, personnel were excluded from the conveyor and collection areas while the atomization process was in operation.

At approximately 7:00 pm a severe explosion occurred, producing a mushroom cloud visible for miles and breaking windows within a half-mile radius. The local Fire Department responded with all available men and equipment, and also requested automatic aid from two other fire departments. However, first-arriving units found only one injured employee who required medical treatment, and a number of minor spot fires.

The injured employee, who suffered second-degree burns on one hip, had been outside of the restricted area. The blast destroyed the bag cyclones, displaced the barrier wall 4–6 ft, damaged the primary cyclones and collection equipment, and ruptured several explosion vents along the conveyor duct. Overall dollar loss was estimated at $40,000.

Investigators believe that the ignition of the aluminum dust occurred in or near the bag house. The extensive measures taken for explosion protection were credited with limiting the loss in this incident.

Fire in Timber Importers, 1970

The fire started in the main machine shop and storage shed when a spark or other foreign body was drawn via extraction ducting into a sawdust cyclone. A smouldering fire was produced and this explosively ignited the dust in the cyclone and ducting. The fire spread rapidly in the timber building and within a few minutes the whole building was afire.

Less than 15 min after the arrival of the Brigade the roof collapsed. Stacks of timber and four small buildings in the near vicinity became involved. One of these was an electricity substation and this was isolated by the Electricity Board. Intense radiant heat was generated and this ignited layers of paint on one of two nearby gas holders, each containing 19,824 m^3 of town's gas. Three plates fractured with the heat and escaping gas ignited. Gas Board engineers lowered the pressure in the holder—a slow operation which took 11 h. It was 13 h after the fire was reported that the leaks in the holder were sealed and the holders declared safe.

The whole of the machine shop and storage shed, a boiler house, and a small shed was destroyed or severely damaged. One wall and quarter of the roof of a box room and the door of the electricity substation were damaged and the two gas holders damaged by radiant heat.

Phenolic Moulding Powder Manufacturers, 1970

A spark from an unknown source ignited phenolic powder dusts in a filter plant with subsequent explosions and fire in filter plants, cyclones, and extractor ducting. A major fire occurred at roof level of the five-storied building in the filter plants and cyclones, but there were only minor fires due to the phenolic powder in machinery on the first, third, and fourth stories and in the extractor ducting. The fire was confined because the explosion that damaged filter plants also damaged the extractor fans, isolating the small fires in the area where they occurred and because several of the machines were not operating at the time of the incident. Although some smoke was produced by phenolic compounds, it did not hamper the work of the Fire Brigade. The structure of the building was not affected.

Sanding of Styrene Butadiene Rubber Sheets, 1978

Sheets of styrene butadiene rubber were sanded by a water-cooled sanding drum whose temperature remained below 60°C. A number of explosions occurred in the bag filter dust-collection system. The bag filters were situated outside the factory and were properly vented; there was therefore no damage other than to the bags themselves. The cause of the ignition was believed to be the very hot dust particles created by the sanding process. Some of these accumulated in the base of the sander, others were swept up by the extraction system. Glowing particles were sometimes observed in the base of the sander. The sanded material decomposed slightly on heating giving some styrene vapor. The dust was therefore probably more explosive than the sample subsequently tested in the laboratory from which no styrene vapor was produced by heating.

6.2.5. Maintenance and Cleaning

A large number of industrial accidents occur during maintenance and cleaning. Often the procedures are less well-defined when the plant is not operating and there is also less thought given to the dangers of residual powder during shutdown periods. If maintenance and cleaning is a deviation from the normal operating procedure the actions of each man involved is less well-defined and it is more probable that the action of one will adversely affect another if procedures are not very strict.

Poor maintenance as well as occasionally being the direct cause of an accident frequently considerably increases damage. Routine inspection in plants handling powders and dusts should include testing vent-opening pressures, servicing suppression equipment, checking alarms and trips, and grounded electrical paths, as well as the normal tests of electrical equipment and fire-fighting apparatus.

Dust deposits must be regularly cleaned from any surfaces inside or outside the plant itself. In the event of an explosion occurring inside a piece of equipment, dust layers outside and in work areas may be raised to a cloud and ignited by hot particles or surfaces from the initial explosion. The secondary explosion can then disperse more dust, resulting in very large explosions and considerable damage.

Plants should be designed to be easy to clean and excess dust should be prevented by ensuring that joints and inspection hatches are leak-tight. Where possible, plants should operate under a slight negative pressure rather than a slight positive pressure. All open ports such as bagging off points or charging chutes should be fitted with dust extraction.

The massive damage and high loss of life in the two grain elevator explosions that occurred in the United States in December 1979 was due to the way in which the explosion was able to spread through a large area of the plant through tunnels and open belt conveyors containing a good deal of dust. In a clean plant a fire or small ignition has a much better chance of being confined to the area where the ignition occurred.

Case Histories—Cleaning and Maintenance

Welding

An explosion occurred during welding of ducting at the top of a grain storage silo. The explosion remained confined to the area and the damage was not great but one man was killed. The accident occurred when two of the elevators were started up in error before the welding was complete.

Blocked Screw Feed

A screw feeder became clogged with powder. The operator loosened the blockage by hitting with a steel bar. The screw became bent and rubbed against the inside of its tube. The powder ignited and an explosion then occurred in the dust bag fitter.

Lighted Match Ignites Dust

While the premises were closed for the annual holidays, a maintenance man carrying out repairs in a metal polishing shop lit a cigarette and carelessly discarded a lighted match. Dust on the floor caught fire and it spread rapidly to the machinery. Three firemen were injured when a dust explosion occurred in the dust-extraction ducting.

Most of the machinery in the section was severely damaged and part of the roof destroyed. Slight damage by fire occurred to a further part of the roof and part of a wall.

Cleaning

A spark from an electrical junction box ignited dust that was being swept away. The damage amounted to £100,000. This clearly illustrates the importance of extending precautions to outside those where dust is supposed to exist.

Cleaning of a Filter in a Dome of a Paddle Drier

An operator had completed mechanical cleaning of a filter in the charging dome of a paddle drier and was flushing it with cold water which ran down into the drier. When he loosened the cover of the dome, the cover was lifted 0.5 m by the internal pressure and the operator sustained first degree burns on his face by the escaping steam.

Prior to the cleaning operation, the steam heating of the drier had been switched off, but its contents were still at a temperature of about 140°C. Thus the incoming flushing water vaporized immediately.

For changing, or cleaning filters the same safety precautions have to be applied as for discharging the plant, i.e., it must be cooled until the temperature falls below 100°C.

Fatal Accident When Cleaning a Vessel

A reaction vessel was to be cleaned for internal maintenance. It was filled with water and heated up to the boiling point with the agitator running. The operator was instructed to cut off the steam to the reactor, let it cool off for 20 min, then blow air through the dip pipe and drain the reactor. However, he opened the air valve much earlier. To reach the air valve, he leaned over the open manhole. The air streaming through the dip pipe forced boiling water out of the manhole. The operator sustained burns over 90% of his body and died several days later. The reactor cleaning procedure was changed.

6.3. THE HUMAN ELEMENT

Reading the case histories of industrial fires and explosions, it can be seen that in a large number of accidents human error either causes the incident or considerably worsens the situation. Surveys have shown that even with well-trained, well-motivated operators performing a sequence of events regularly errors will be made. Attempts have been made to quantify this, for example, Kletz (1980) suggests that in a busy control room an error might be made, 1 in 10 times if rapid action is required. This might be reduced to 1 in 100 if the room is quiet. In a simple familiar startup routine a man might forget one particular step 1 in 1000 times. Even these statistics are far too frequent if the result of the wrong action could lead to hazardous conditions. The assumption must be made that errors will occur and processes should either be made automatic to remove the human element or a system of alarms and checks should be installed so that the fault situation is discovered and remedied. The choice between these two courses of action obviously depends on the seriousness of the result of the error and the types of check available.

The position is even worse in situations where education or motivation is less good. If an operation is more easily carried out one way than another, short cuts will be taken regardless of regulations to the contrary. The manual placing of ground clips onto containers before filling is a particular case where regulations are frequently disobeyed. It seems that this is not simply a case of the operator occasionally forgetting the clip, but a lack of belief in the danger of omitting it, perhaps because lack of attachment does not instantly result in a fire.

The human element does not always operate on the negative side. An alert operator can frequently observe signs of problems developing before they have reached hazardous proportions. For example, smouldering dust, slipping conveyor belts, and overheating machinery usually smell. Signs of static electricity may be noticed and reported as well as any indications of burnt or charred particles. It is important that the workforce is educated in the hazards of dust explosions and this is particularly a problem in plants producing foods or common substances not normally associated with hazards.

6.4. DUST CONTROL

Fine dusts tend to be produced by most powders during processing. Dust-control systems must be designed both to prevent or reduce the concentration of explosive clouds within the system and to prevent the escape of dusts from the system in order to eliminate potentially hazardous layers of deposits. Two means are available. The dust may be controlled and collected, or processes may be designed to reduce the amount of dust formed in the first place. Many processes can be modified so that attrition, air entrainment, and unnecessary impaction of powders

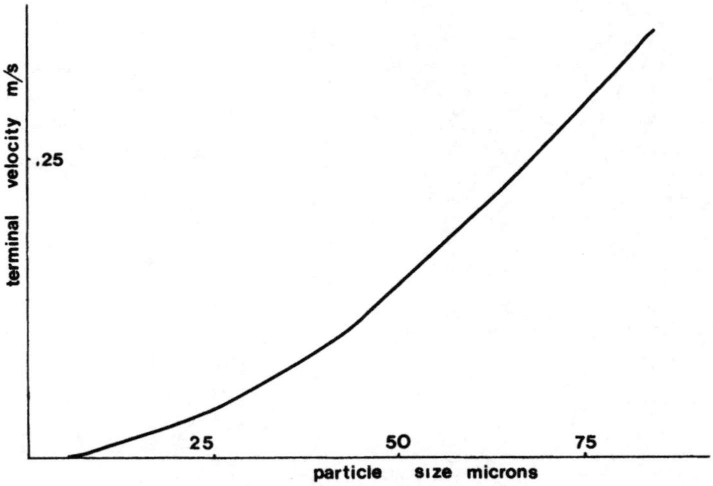

FIGURE 6.4. Terminal velocity of dust particles.

are avoided. This not only improves safety but often also improves the quality of the product.

Dust is created by impact and friction of bulk powders and separated from the bulk powder because of its different behavior when it falls under gravity or is subjected to an air stream. The falling velocity of particles as a function of powder diameter is illustrated in Figure 6.4. It can be seen that the settling rate of a fine dust is often the same order as natural air currents and even relatively coarse particles will remain in suspension under the influence of typical industrial air currents. For example, convection currents may be of the order of 0.3 m/s and particles below 20 μm in diameter may therefore remain suspended in the air.

Within a processing plant dust can usually be successfully confined by means of ducting. Clouds inside systems can be reduced by decreasing the distance over which powder falls or using a sloping chute, but in many systems the only practical method of reducing clouds within the plant is an effective extraction system. A dust-extraction system can be designed to have a sufficiently high air output so that the dust concentration within the extraction ducting is below the minimum explosive concentration. However, extraction systems must always be treated as a potentially hazardous part of the plant, since fault conditions that lead to an overloading of dust or a reduction of the air could produce a potentially hazardous situation.

The escape of dust from the system to form layers and deposits outside the plant occurs primarily at loading and unloading points,

(a)

(b)

FIGURE 6.5. A tripper on a grain elevator (a) with and (b) without effective extraction.

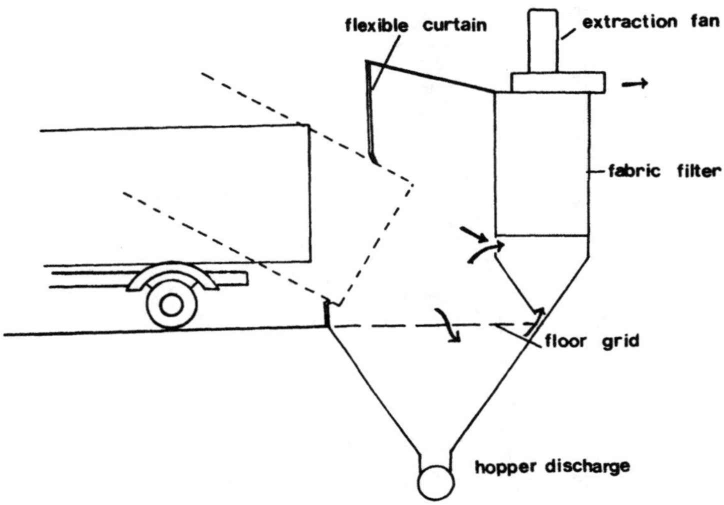

FIGURE 6.6. Dust control for lorry unloading.

although other positions where the material is transported in uncovered systems such as open conveyor belts will also produce dust clouds (Figure 6.5).

Cleaning the dust once it has escaped is both time consuming and difficult to enforce, and it is far more effective to prevent it from leaving the system. A typical system for containing dust during delivery of bulk powder from an open lorry—a system still used for some low-cost bulk-volume materials—is shown in Figure 6.6.

Higher-cost or lower-volume materials are often delivered in sacks or bags and manual tipping of these can create a considerable dust problem which is not only an explosion hazard but also a danger to health. Bag-slitting and emptying machines now exist that incorporate a dust-control system and dispose of the emptied sacks. Although these machines successfully control dust emission, some contain fast-turning cutting blades to rip open bags which may themselves provide an ignition source. Bag-slitting machines must therefore be protected from explosion by suitable venting, inertion, or suppression.

In some processes dust can be controlled by choosing a system in which water is used to damp down the dust. A fine water supply may be used to eliminate hazardous dust clouds or a water-based system may be chosen in preference to a dry system. For example, wet scrubbers are less hazardous than other types of dust-collection equipment.

The use of a fine charged water mist for reduction of dust has been studied by Hoenig of the University of Arizona (1978). The majority of

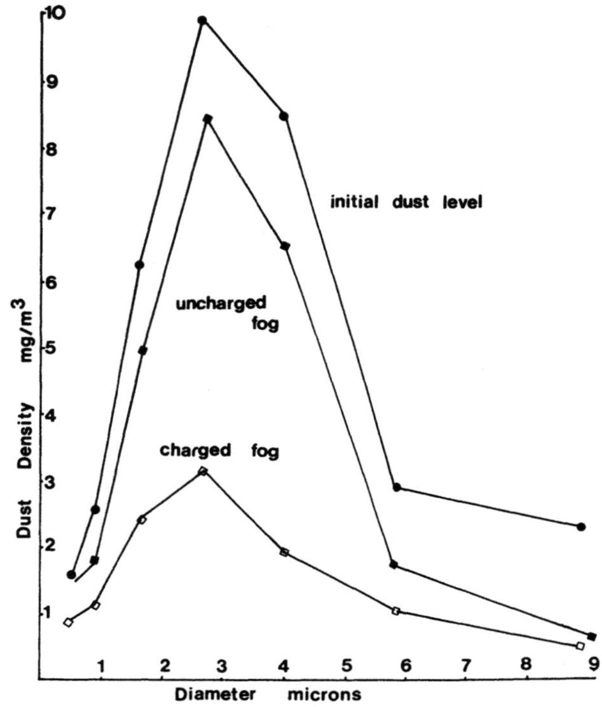

FIGURE 6.7. Reduction of dust by a charged fog.

fine dusts are electrically charged and Hoenig shows that the dust density is considerably reduced—particularly for the respirable dusts less than 10 μm in size—if the dust is sprayed with a mist that is intentionally charged to the opposite polarity. Figure 6.7 shows the reduction in dust density for foundry dust using a charged and uncharged fog. The water droplets in the fog can be extremely small so the total increase in water content is not large. At present this technique is not in common usage but the preliminary research shows considerable promise.

7

Dust Explosion Protection

7.1. INTRODUCTION

It is impossible to be certain that all dust ignition sources have been completely eliminated. Human error, equipment failure, and procedural errors will occur, and plant design must always assume that fires and explosions can and will occur in any region where there is a flammable dust. Precautions must be taken to minimize damage and injury in the event of fire in any area where there is combustible material, and the possibility of a dust explosion must be considered not only in areas where a cloud of explosive concentration is expected, such as in storage bins, grinders, driers, and dust-collection units, but also in areas where there are layers or heaps of dust which might be raised to a cloud by an explosion in another area or an unexpected air stream.

Basically there are four methods of minimizing damage in the event of a dust explosion:

(i) The vessel or enclosure may be built strong enough to withstand the maximum pressure generated in the ignition, so that the explosion is contained within the chamber in which it occurs.

(ii) The vessel may be vented so that the pressure created in the explosion opens an explosion door or bursts a diaphragm of weaker material and the rise in pressure in the vessel is reduced. It should be noted that since the pressure in the vessel is rising rapidly as the vent opens the maximum pressure in the vented vessel is always significantly greater than the pressure at which the vent opens and the size of this overpressure depends on the rate at which the pressure rises and the size of the vent.

(iii) Automatic suppression may be fitted so that the initial stages of the pressure rise during the explosion trigger the release of an extinguishing medium which prevents further pressure rise. Suppression still involves a pressure rise in the vessel from the level at which the system triggers, but there is a considerable reduction in the maximum pressure achieved during the explosion.

(iv) The system may be rendered inert so that there is insufficient oxygen present to sustain combustion.

In addition, precautions must be taken to prevent the spread of fires and explosions from one part of the plant to another. This may involve the use of flame traps and barriers, and automatic valves, and also good housekeeping and plant cleanliness so dust outside the plant and between different parts of the system is minimized. The design of the plant is very important here, both for ease of cleaning and containment of dust and from the point of view of the siting and design of general dust systems such as dust-extraction systems.

Although the greatest damage is done in situations where a dust explosion occurs many dust explosions start as small fires. Fire protection systems are therefore an essential part of explosion protection. Several of the case histories that have been cited in earlier chapters involved dust explosions which developed from fires not satisfactorily extinguished either by the fire extinguishers available or by the automatic sprinkler systems.

7.2. CONTAINMENT OF EXPLOSIONS

The maximum pressure developed in an explosion can be measured in a 1-m^3 vessel described in Chapter 5. The pressure developed is not a strong function of vessel size and pressure values measured in this apparatus may be assumed to apply directly to a larger plant. In fact it is unlikely that the dust concentration will be optimum, therefore, the 1-m^3 values should allow a reasonable safety margin. It should be noted that the maximum pressure measured in the Hartman bomb is often 2–3 times lower than that measured in the 1-m^3 vessel. It is therefore safer to use the 1-m^3 values. These are published in *Forschungsberericht Staub Explosion*.*

At low dust concentrations the pressure developed in an explosion may be barely detectable but, under optimum conditions of concentration and dispersion, pressures range typically from around 5–6 bars for dusts such as sugar, flour, and epoxy to 13 bars for aluminum dust.

Propane has a maximum pressure of 7.9 bars and methane 7.4 bars. When a hybrid mixture of a dust and a vapor is formed the maximum

*ISS NNr 0172 729X.

pressure achieved tends to be slightly higher than that produced by either component separately. Thus propane has a maximum explosive pressure of 7.9 bars and a sample of polyethylene of 80-μm median diameter has a maximum pressure of the order of 7.5 bars, but the maximum pressure rise at an optimum propane content of air of 3% by volume was just over 9 bars.

It can be seen that containment involves design of very substantial vessels produced according to the design codes that have been developed for pressure vessels. This becomes progressively more difficult the larger the vessel considered, and is frequently impractical for large plants.

In some cases where explosions are rare and other protection methods unsuitable, it may be possible to build a *pressure-shock resistant vessel* which has a considerably lower mechanical strength. This vessel is designed to withstand the maximum pressure without rupture but it is accepted that in the event of an internal explosion there may be permanent deformation, and extensive repairs or replacement may be necessary.

Containment of explosion is most frequently chosen as protection for a small plant such as one containing a few grinding machines. Pipelines are also frequently designed to withstand the maximum pressure of an explosion because their geometry makes this a feasible proposition; venting of a long line is difficult and may serve to accelerate the flame, increasing the risk that a deflagration will become a detonation with very much higher peak pressures.

Approved codes generally specify that pipelines and ducts in which an explosion may occur should be built to withstand a pressure of 10 bars. This will in fact withstand the extremely short peak pressure of a detonation on straight sections of pipe although it is possible that damage would occur on bends and flanges to branch pipes.

7.3. EXPLOSION VENTING—PRESSURE RELIEF FOR EXPLOSION PROTECTION

7.3.1. Introduction

Explosion or *combustion venting* is a means by which the maximum pressure due to a deflagration in an enclosure may be limited to a level that will not cause excessive stress or damage to the enclosure structure. This is accomplished by release of the products of combustion to the exterior of the enclosure by means of an orifice or vent. This may be an open or hinged door or a bursting disk of a weak material which is ruptured by the pressure rise of the explosion. The design of the venting system must take into consideration the dimensions, volume, and construction of the enclosure and the deflagration properties of a combustible material.

Explosion venting is a protective measure to reduce the effects of an explosion and it does not remove the need for other measures of explosion prevention. The purpose of the vent is to limit the maximum pressure resulting from a deflagration in order to prevent damage to the enclosure. Extensive damage may result if combustion occurs within an enclosure too weak to withstand the force of the deflagration. A wall of 8-in brick or 8-in concrete block will not withstand a sustained internal pressure as small as 1 psi (6.9 kPa). In venting enclosures the maximum pressure developed should not exceed that which the enclosure can safely withstand.

Explosion venting will only give full protection against a deflagration and not the shock wave associated with a detonation, which occurs when the flame front can accelerate and reach the velocity of sound. In addition, it must be borne in mind that considerable care must be exercised in the design, installation, location, and maintenance of a venting system if any explosion is to be successfully released and if secondary fire and explosion hazards are to be avoided.

In particular, it is important that the venting system is capable of ejecting the deflagration products clear of the enclosure in a manner that does not present a secondary hazard to buildings, plant, or personnel. It often happens that the initial deflagration is limited to the area associated with the ignition source within the enclosure. However, the pressure rise

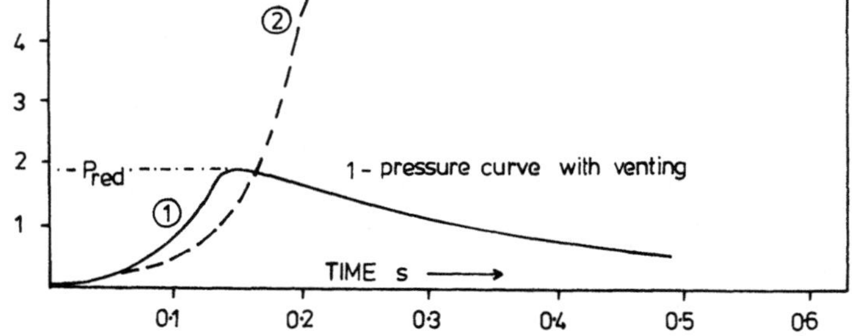

FIGURE 7.1. Pressure–time curve for a vessel with and without explosion venting (propane/air) (Bartknecht, 1971).

can eject a plume of unburned material through the vents and this can travel a considerable distance, perhaps up to 30 m or more. The flame front can propagate through the ejected dust plume, igniting it outside the enclosure area and creating a considerable secondary hazard.

An illustration of the level of protection that can be gained from explosion venting by reducing the maximum pressure of the explosion to below that which the enclosure can safely withstand is given in Figure 7.1. It shows the pressure–time curve for a vented and nonvented dust explosion in a 1-m^3 test chamber (Donat, 1977). Instead of a maximum explosion pressure of approximately 6.8-bars gauge in gas/air explosions (or 7.8–12.6 bars gauge in dust/air explosion), only a reduced explosion pressure of about 0.17–1.93 bars gauge occurs in the apparatus with explosion venting.

7.3.2. Advantages and Disadvantages of Explosion Venting

Pressure relief (explosion venting) has the following advantages over other methods of explosion protection:

(i) Capital expenditure is required only once; that is, there are no operational expenses as are incurred with inerting or an active explosion suppression system.

(ii) Maintenance of venting systems is less expensive (but no less essential) than the maintenance of inerting or explosion suppression systems.

(iii) The design of the vessel regarding pressure resistance does not have to correspond to the maximum explosion pressure. It is sufficient for the enclosure to be designed to withstand a much reduced pressure (P_{red} in Figure 7.1).

There are, however, certain disadvantages associated with explosion venting:

(i) Where the enclosure containing the combustible material is situated within a building, a vent duct must be provided in order to exhaust the combustion products and unburned material to the exterior of the building. This reduces the efficiency of the vent which must be correspondingly increased in size. The area or zone into which flame and material might be ejected must be secure so that it does not seriously endanger personnel or cause other secondary hazards. Vertical ejection might prove safer than horizontal but the siting of vent ducts within a building can prove a problem and the ducts themselves may increase the hazard. This is discussed in Section 7.3.9.

(ii) Containers or enclosures situated outdoors do not normally require ducting, but in these cases it must be remembered that the vented plume of combustion products may extend to greater than 30 m. The burning powder cloud or particles may be carried considerable distances by the wind.

(iii) Ineffective venting can easily arise due to lack of maintenance or servicing of venting systems. It is all too easy to assume that a *passive* simple mechanical device such as a hinged vent or bursting disk will take care of itself. *Active* systems of explosion protection require regular attention either because they are in regular operation (such as an inerting system using carbon dioxide or nitrogen) or because they are regularly serviced by the company that installed the system (such as an active suppression system triggered by pressure rise due to a deflagration). Corrosion, abrasion or material fatigue can lead to failure of a venting system. In addition, careless painting or even freezing of weather-exposed vents can prevent the system functioning, with the result that it offers either only partial protection or none at all in the event of an explosion. Regular inspection of such explosion protection devices are therefore of paramount importance.

(iv) Venting may only be used where the product can be allowed to escape into the atmosphere. Toxic materials cannot be vented.

(v) If an explosion does occur and is effectively vented there is no permanent damage to the plant. However, the product is lost and extensive cleaning is often required. Production is disrupted. Venting is therefore seldom a satisfactory solution where it is difficult to remove ignition sources and ignitions are common.

(vi) Effective venting requires that an orifice of sufficient area be placed in a suitable position in the direction of the blase wave of an explosion. Effectively this means in the roof of a cylindrical vessel. If the plant has not been designed with venting in view, it may not be possible to find room for a vent of sufficient area for the volume and design strength of the vessel. In this case other precautions must be taken. Pipes and ducts require frequent vents along their length and this may be impractical.

7.3.3. Considerations in the Use of Explosion Venting

There are several questions that need to be answered once the design engineer has decided upon explosion venting as a protective system.

(a) What is the best location for the relief system to be installed?

(b) Does the vent or bursting panel need ducting to the exterior of the building and if so of what dimensions should it be and where is it to lead?

(c) Should a bursting disk or panel be chosen for pressure relief or an explosion door?

(d) How should the vessel be designed, built, and tested in order to ascertain that it can adequately withstand the reduced explosion pressure experienced during explosion venting?

7.3.4. Explosion Pressure Development in a Small Container

For the combustion of a gas or powder a number of conditions have to be fulfilled simultaneously: (i) A combustible material or gas must be present. (ii) An oxidizing agent must be present, usually air. (iii) The combustible mixture must be within the explosive limits, i.e., neither too rich nor too lean for combustion to take place. (iv) An ignition source must be present that has sufficient energy to ignite the explosive mixture.

In the case of an explosion the maximum explosion pressure P_{max} and the maximum rate of pressure rise $(dp/dt)_{max}$ define the comparative violence of the explosion. Figure 7.2 illustrates a pressure–time curve for the explosion of a propane/air mixture in a 1-m³ vessel. Figure 7.3 shows the explosion pressure of polyethylene measured in the Hartman bomb. The maximum pressure P_{max} is defined as the maximum peak pressure. The maximum rate of pressure rise $(dp/dt)_{max}$ is defined as the maximum slope of the curve and the average rate of pressure rise as the slope of the line from $t = 0$ to t at (P_{max}). It is also possible to define an induction time which is the time between the activation of the ignition source and the

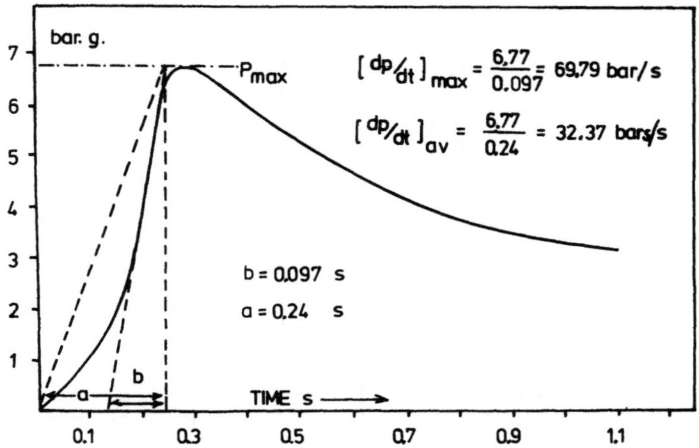

FIGURE 7.2. Pressure–time curve for a propane/air mixture in a closed 1-m³ vessel (Bartknecht, 1971).

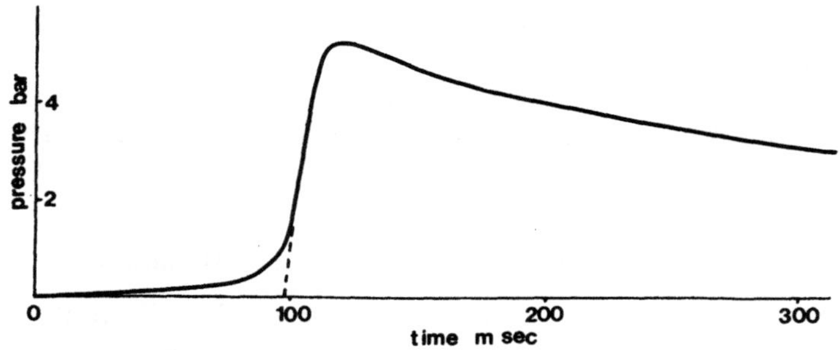

FIGURE 7.3. Pressure during an explosion of high-density polyethylene.

onset of the explosion. This may be defined either as the time between activation of the ignition source and the intersection of the tangent with maximum slope with the abscissa as shown in Figure 7.3, or the time between activation and an arbitrary defined pressure change.

The values of maximum pressure rise P_{max} and maximum rate of pressure rise $(dp/dt)_{max}$ for a given system will depend on the concentration of the combustible mixture. The concentrations that must be considered are those for which P_{max} and $(dp/dt)_{max}$ attain their highest value for a given material. Figure 7.4 illustrates the variation of P_{max} and $(dp/dt)_{max}$ for various propane/air compositions. It can been seen that both P_{max} and $(dp/dt)_{max}$ peak at 4% volume of propane in air.

The maximum explosion pressure P_{max} is largely independent of the size of the containing space. However, the maximum rate of pressure rise

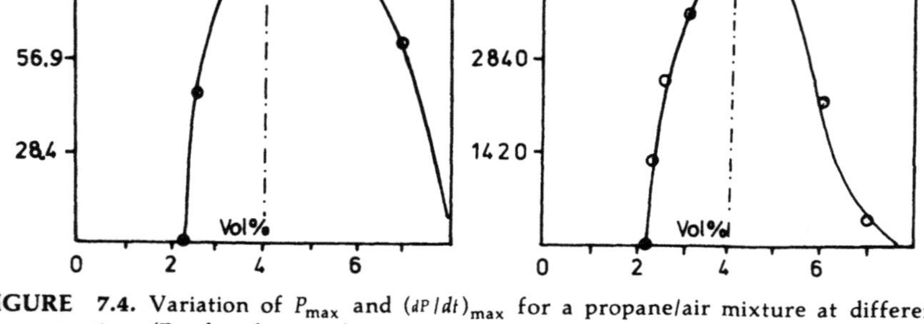

FIGURE 7.4. Variation of P_{max} and $(dP/dt)_{max}$ for a propane/air mixture at different concentrations (Bartknecht, 1971).

$(dp/dt)_{max}$ is largely influenced by the size of the enclosure. For a given system the maximum rate of pressure rise becomes smaller as the volume of the vessel is increased.

The relationship obtained between $(dp/dt)_{max}$ and vessel volume is known as the *Cubic Law* (Maisey, 1965) and is valid for all practical purposes. The Cubic Law may be stated as

$$\left(\frac{dP}{dt}\right)_{max} V^{1/3} = K \quad \text{(a constant)} \tag{7.1}$$

The constant K has the units bar · m/s and is expressed as K_G for gases and K_{st} for powders or dusts. Typical values of K_{st} for some common materials were given in Table 5.5. The values in this table are the maximum values obtained with these dusts in a 1-m³ chamber with optimum dust concentration.

Particle or grain size distribution also has a considerable effect on explosibility (Hartman *et al.*, 1950). In the determination of explosion data this must be taken into account. Fine dusts, as might be expected, explode more violently than coarse dusts. For example, a dust with a mean particle size of 10 μm will explode more violently than a dust with a mean particle size of 300 μm.

An additional factor which must be taken into consideration is the degree of turbulence of the combustible mixture. It is found that a different delay in the ignition time (the time between dispersion of the dust and striking of the ignition source) results in different values of P_{max} and $(dp/dt)_{max}$ due to differences in turbulence. Turbulent mixtures being more explosive, the worst possible case must be ascertained in the determination of $(dp/dt)_{max}$ (Moore, 1979).

7.3.5. Application of the Cubic Law

Tests in large enclosures show that the Cubic Law, which is known to be true for closed, unvented vessels, also applies within limits to vented vessels. The basic equation is

$$\frac{f_2}{f_1} = \frac{(V_1)^{1/3}}{(V_2)^{1/3}} \tag{7.2}$$

Here f_1 is the vent ratio (vent area/vessel volume) obtained for a test vessel of volume V_1. f_2 and V_2 correspond to the required vent ratio and volume of the vessel in question. The vent area S_2 in m² required for the vessel is then

$$S_2 = S_1 \left(\frac{V_2}{V_1}\right)^{2/3} \tag{7.3}$$

Vessels of different volumes require different vent areas to reduce the maximum explosion pressure to a given level. However, it is not necessary to increase the venting area determined for a small vessel by an amount directly proportional to the volume of the larger vessel. The larger the vessel volume the smaller is the required vent ratio to obtain a particular reduced pressure.

Information on how the reduced explosion pressure varies with the vent ratio f and size of the enclosure is used to construct nomograms of the type shown in Figure 7.5. The nomograms can be used to determine the required venting area for a given situation defined by the vessel volume, the K_{st} value of the dust (or K_G value of a gas), the maximum safety allowable pressure in the vessel during venting, and the pressure in which the vent initially opens.

K_{st} values obtained for a particular dust can depend on the test apparatus and test method used. The Hartman apparatus as described in Chapter 5 for assessing dust explosibility has been the accepted means of testing dusts in the U.K. (Dorsett *et al.*, 1960) and has served as a basis for the design of venting and explosion measures. The suitability of the Hartman apparatus as a means of categorizing dust explosibility has been questioned by Bartknecht. The small volume and shape of the test apparatus leads to an under estimate of $(dp/dt)_{max}$ by a factor of 2–3 and extrapolation to larger vessels is unreliable. Turbulence, ignition delay, and ignition source can also have a significant effect when comparison is

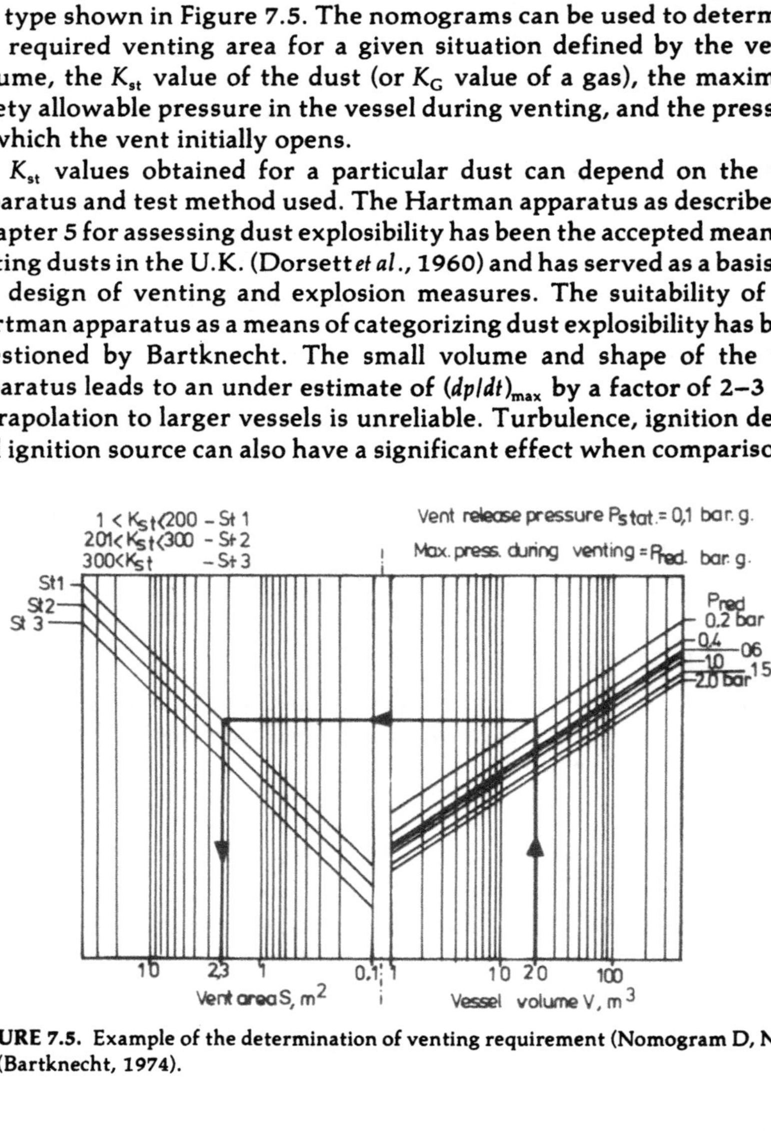

FIGURE 7.5. Example of the determination of venting requirement (Nomogram D, NFPA 68) (Bartknecht, 1974).

made between one apparatus and another. A comparison of different test methods is given by Moore (1979) and a detailed description of a similar test method to the Hartman is given by Schwab and Othmer (1964). Full details of the Hartman and other test methods is given in a Bureau of Mines publication Report 5624 and in Chapter 5 of this book.

For vent design the dust is classified broadly into three groups: St 1 with K_{st} up to 200, St 2 with K_{st} between 201 and 300, and St 3 where K_{st} exceeds 300. For gases the value of K_G is dependent on the energy of the ignition source. This does not appear to be the case for dusts, and a weak spark discharge gives the same K_{st} value as a pyrotechnic igniter of 10,000 J. The American classification system, however, considers weak and strong ignition sources separately in analogy with the treatment for gases. For a low-energy ignition source the American system classifies St 1 of K_{st} less than 100, St 2 between 101 and 200, and St 3 above 300. The European system considers all dust ignitions as if they were strong ignition sources. A typical nomogram is illustrated in Figure 7.5. This is nomogram D (NFPA Guidelines) which applies to venting of flammable dusts in the low-pressure range. Other nomograms are listed in the guidelines for a range of gas compositions, vent release pressures, and ignition sources.

As an example of the use of these nomograms we can calculate the vent area required for a 20-m^3 vessel. In Figure 7.5 the volume of the enclosure or vessel is shown on the right-hand half of the x axis. The vent area required is shown on the opposite half of the nomogram. The three ranges of K_{st} for dusts are also indicated on this half of the nomogram. This nomogram assumes a strong ignition source. If our dust has a K_{st} in the range 1–200 bars · m/s we use the St 1 line on the nomogram. Assume we require the vent release pressure to be 0.1 bar and the maximum pressure during venting to be equal to 0.2 bar. The requisite vent area is determined by selecting the vessel volume on the $V(m^3)$ axis and vertically following the value 20 m^3) up to the 0.2-bar line (maximum pressure during venting). Having reached the 0.2-bar line we traverse across maintaining this value to the St 1 line on the left-hand side of the nomogram. At the intersection with the St 1 line we drop vertically down to the vent area axis (vent area S, m^2) where the required vent area if 2.3 m^3.

7.3.6. Limitation of the Use of the Cubic Law

The cubic law on which the nomogram is based has been extensively verified for volumes up to 60 m^3. Work on larger vessels shows that as the vessel size increases the cubic law tends to overestimate the necessary vent area.

The cubic law applies only to vessels whose length-to-diameter ratio is less than 5:1. If the vessel is more elongated than this, a strong axial flow develops and the development of the pressure wave is more similar to an explosion in a pipeline than in a spherical test vessel. In this situation the whole of the roof area should be vented and this area must still exceed that given by use of the nomogram.

The pressure developed during an explosion is directly proportional to the initial pressure in the vessel. Thus if the pressure in the vessel is decreased below atmospheric the results will be correspondingly reduced, but if the pressure should be increased from atmospheric (1 bar) to 1.5 bar, both the maximum pressure and rate of pressure rise will also be increased by a factor of 1.5. The consequence of this is that if two vessels are joined together and an explosion in the first occurs, the pressure pulse will reach the second vessel ahead of the flame front. The second vessel therefore ignites at elevated pressure with consequently higher pressures and rates of pressure rise produced. Vent calculations should allow for this effect unless isolation valves are fitted.

7.3.7. Vent Design

Having assessed the requisite vent area, the type of vent to be installed needs to be considered. It must be as light as possible for maximum effect during a deflagration, preferably not above 6–7 kg/m^2 (10 kg/m^2 maximum). It must also be borne in mind that counterbalanced panels of greater mass-to-area ratio than the above will not give the requisite venting performance due to their high inertia and this will result in an increase in the maximum pressure generated during venting.

Open or unrestricted vents are most effective in relieving pressure buildup. Vents covered with a diaphragm, swinging doors, and bursting disks all require some inertia to be overcome and are correspondingly less effective.

There is a considerable body of literature on vent design; for example, NFPA (1978), Bartknecht (1981), Donat (1977), Nagy *et al.*, (1950), Palmer (1971, 1973), and Cocks (1979). A more extensive literature list is given in NFPA (1978).

7.3.7.1. *Bursting Disks*

Bursting disks are either made of a material which when activated disintegrates into granules or of a lightweight diaphragm which splits and blows away. Most of the granule type of bursting disks are made of resin-impregnated graphite. They activate at a pressure exceeding 0.1 bar. The membrane-type may be aluminum foil, plastic, or a composite paper-type of material. Bursting disks have the following advantages over solid vents: (a) low manufacturing cost; (b) they can easily be made air tight; (c) they have low weight, therefore they have little effect on the maximum overpressure in the chamber; and (d) it is less easy for the disk to fail since there are no mechanical catches or hinges that may seize.

However, the mechanical strength of the disk is low and this may be a problem on large-area containers. It is possible to increase the mechanical strength by placing an open grid under the disk (this should never be over the disk where it will restrict opening). Bursting disks also have to be replaced periodically because of deterioration due to weather and fluctuations of pressure in the plant.

In order to calculate the vent area, it is important that the pressure at which the disk will burst is known. The static bursting pressure may be calculated approximately from the equation

$$P_{\text{stat}} = \frac{d\ \delta B}{D} \tag{7.4}$$

where P_{stat} is the static pressure, d is the thickness of the disk, D is the diameter of the aperture cover, and δB is the tensile strength of the material. However, variations in material require an experimental test of static opening pressure. The way in which the disk is clamped can affect the opening pressure; therefore this must be simulated in experimental trials.

The pressure at which the disk bursts will also depend on the rate of pressure rise. The difference between the static pressure at which the disk bursts and the higher dynamic pressure when the pressure is increasing rapidly is greatest for disks that already have a high bursting strength and also greater for smaller appertures.

Materials that stretch easily, such as polyethylene, may bow considerably before splitting. A bursting disk may be designed so that distortion brings the disk into contact with a cutting tool. This reduces the time before the disk bursts and hence reduces the overpressure. It is also possible to have a bursting disk that is ruptured by a small detonator which is activated by a signal from a pressure detector. This type of bursting disk is normally made of special glass. The bursting disk may in any case be crossed by a fine wire carrying an electric current. The bursting of the disk breaks the copper wire and interrupts the current. This can be used to activate automatic valves or shut off other parts of the process.

7.3.7.2. Trap-Door Vents

Solid vents have the major advantage that they can be automatically closed again after the explosion. This decreases the time required to set the plant back in operation and also limits the escape of burning and combustible material. The access of oxygen to burning material left in the vessel is also limited. However, maintenance is absolutely essential to ensure that doors always open at the designed pressure.

Although solid vents are lightweight they can be a considerable danger if they become missile during the explosion. Various retaining devices have been designed and systems exist which dampen the impact of the opening of the doors so that hinges are not damaged and the reclosure of the doors is accelerated.

Vent catches may be spring clips, easily broken retaining clamps, or magnetic catches. The magnetic type of catch can be adjustable so the vent-opening pressure can be controlled. The magnets themselves are fitted to the surroundings of the vent cover rather than to the cover itself, because of the increase in weight they would give.

Specific vent-door designs are described by Palmer (1973) and in the NFPA guide to venting (1978). Provided that the door opens to an angle of at least 45° it has been shown that partial opening of the vent cover does not inhibit venting. The shape of the vent opening is also not critical.

Trap-door vents must operate in all weather conditions and must not freeze shut in cold weather. It is also possible for high winds to exert pressures of up to 2 bars and this can inhibit opening of a poorly designed vent.

7.3.8. Venting of Buildings

The purpose of venting a building is to prevent serious structural damage and the production of lethal projectiles. The strength of all parts of the building structure must be considered including windows, ceilings, walls, roof, and floors.

Most buildings can withstand only very low pressures from within—floors, for example, are not designed to withstand an upward force. The vent area is calculated from the Runes equation (Runes, 1972). This is

$$A_v = \frac{Cl_1 l_2}{p^{1/2}}$$

where A_v is the vent area, C is a constant depending on the burning velocity of the material, p is the pressure that can be withstood by the weakest member, l_1 is the smallest dimension of the building, and l_2 is the second-smallest dimension of the building. C depends on the exploding material as shown in Table 7.1.

The vents or weak panels in buildings should be distributed over the walls of the building rather than all in one wall. The material should not

TABLE 7.1. Values of *C* for the Runes Equation.

Exploding material	C (units SI)
Hydrogen	17.0
Ethylene	10.5
Organic dusts	6.8
Fine metal dusts	10.5

break into sharp or dangerous pieces and should be less than 7.3 kg/m^2 in weight. Vents which will be weak enough to open in the event of an explosion will not be strong enough to support the weight of a person falling against it and therefore should be protected by some method that does not obstruct the opening.

An extensive series of full-scale tests on different systems of venting buildings has been reported by Howard and Karabinis (1980). Different types of vent panels, fastners, and retaining straps were used and the effects of gas explosions inside the test building were filmed. The following guidelines were offered.

(i) Vent panels should be made of a ductile material—brittle materials tend to shatter and form dangerous missiles.

(ii) The weight of the panel should be less than 10.5 kg/m^2.

(iii) Relieving fastners should be selected by testing, using a method that simulates the stress of the explosion. Two types of screw and a rivet were tested by Howard and Karabinis (1980) and the rivet was found to free the panel most reliably.

(iv) The panels must release at as low a pressure as possible during an explosion but must withstand high winds. A release pressure of 0.014–0.019 bar is normally suitable.

(v) A restraint system must be fitted to the panels to prevent them from becoming projectiles. This must be very strongly fixed to the building. Shock-absorber systems lessened the tendency for the restraint system to fail.

It was found that the recommended *C* values in the Runes equation gave a good safety margin and peak pressures measured in the building were generally below 1 psi (0.07 bar) with *C* values of 2 (in S.I. units).

7.3.9. Pipes and Ducting

7.3.9.1. Detonation

In a pipeline or duct the flame front may be able to accelerate so that the velocity exceeds the velocity of sound. The deflagration then becomes a detonation, and the resultant shock wave can produce very high peak pressures, reaching instantaneous values of 25–30 bars. The

TABLE 7.2. Pipe Length for Acceleration to Detonation.[a]

	Pipe length in m for detonation		
Pipe diameter, mm	10.0	200	400
Flammable gas			
Methane	12.5	18.5	>30
Propane	12.5	17.5	22.5
Hydrogen	7.5	12.5	12.5

[a] Data from Bartknecht (1981).

behavior of a flame front in a pipeline attached to a vessel that contains a powder cloud is only slightly affected by whether the pipeline contains powder or is empty; therefore the following considerations apply both to pipes attached to chambers containing a flammable mixture, such as vent ducts and air extraction pipes, as well as ducts containing powder such as in pneumatic lines and flash driers.

The tendency of pipelines and ducts to accelerate flame fronts has been studied for gases by Peterson (1967) Pineau, *et al.* (1978), and Bartknecht (1981).

The flame velocity is greatest where ignition occurs at the closed end of a pipe and the other end is open. The flame speed increases with pipe length. Below a critical pipe diameter acceleration will not occur. This is generally a few millimeters for a gas such as propane but a few centimeters for a powder. In a turbulent gas/air mixture all flammable gases can be forced to undergo detonation. The pipe length required for detonation decreases as the pipe diameter decreases (Table 7.2). When the gaseous mixture is already moving at the time of ignition shorter lengths may be sufficient to give detonation. Pipelines of 10–20 m are generally sufficient to accelerate the flame front of a gaseous explosion to detonation, whereas for flammable dusts with K_{St} >200, 20–40 m is required. The development of detonation is more likely for small pipelines than large.

7.3.9.2. Protection of Ducts and Pipelines

During a detonation, peak pressures may reach 25–30 bars, however, pipe lines designed to withstand 10 bars will generally not be damaged by the very short peak pressure of a detonation. However, lightweight riveted constructions will not survive. Effective venting of pipelines and ducts is only possible if vents are installed over the entire length at frequent intervals (approximately 2 m). In the event of explosion a flame is likely to issue from each opening. This is unlikely to be a practical solution to the problem of explosions in pipelines except in the open air, and containment, suppression, or automatic valves are usually used.

7.3.9.3. Vent Ducts

It can be seen from the previous sections that the necessity of introducing a duct between the opening of a chamber to be vented and the outside introduces a number of problems.

Although ducts up to 20 m in length are not likely to produce detonations (particularly at the high duct diameters used for venting) these will still allow acceleration of the flame front and increase in pressure. A duct always reduces the effectiveness of a vent and the duct length must be compensated by an increase in vent area. The United States Bureau of Mines has carried out tests in a 0.028-m^3 chamber with ducts varying between 0.08 and 0.015 m in diameter. A 3-m duct increased the maximum overpressure by a factor of 10. A 45° bend increased pressure by 30% and a 90° bend was approximately equal in effect to a straight length of pipe 2–3 times the length of the bend.

Donat (1977) quotes the example of a particular vessel in which a pressure of 0.2 bar was increased to 0.7 bar by the introduction of a duct between 1 and 3 m long. A duct exceeding 3 m in length increased the maximum pressure to 1.7 bar. Figures 7.6 and 7.7 show results of the effects of vent ducts published by Pineau, *et al.* (1978) and by Bartknecht (1981).

If ducts must be used it is necessary to design both the chamber and the duct itself to withstand the much higher maximum pressure produced with a ducted vent. If absolutely necessary, ducts should be

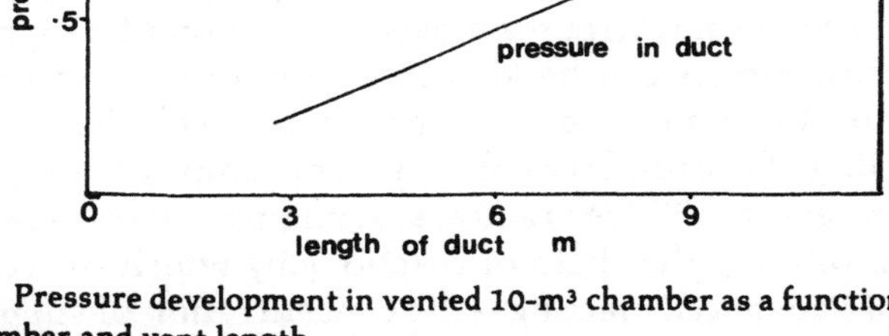

FIGURE 7.6. Pressure development in vented 10-m^3 chamber as a function of dust length between chamber and vent length.

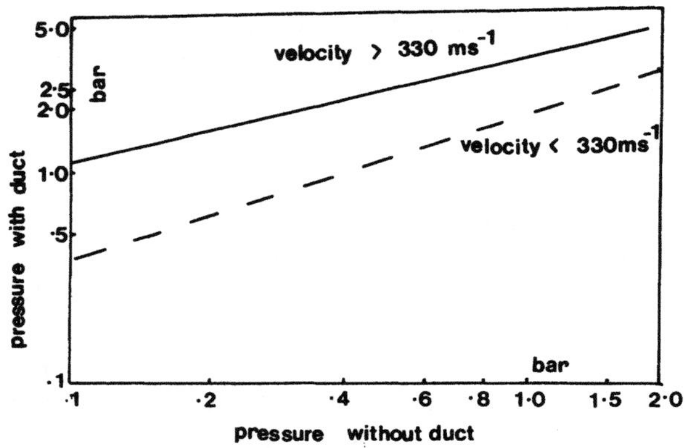

FIGURE 7.7. Effect of duct on pressure in a vented chamber.

short and straight, and their area should exceed the required vent area. British and American regulations specify a maximum duct length of 3 m beyond which the duct must be strengthened to withstand pressures of 10 bars.

7.4. EXPLOSION SUPPRESSION

7.4.1. Introduction

The philosophy of explosion suppression is to detect the explosion very early, before it has time to develop, and to blanket the system with an extinguishing medium which prevents the explosion from developing further. The time between the first signs of a pressure rise or flame and the peak pressure pulse for many dusts is between 0.01 and 0.07 s. Detection and initiation of the release of a suppression takes on the order of ~0.002 s; therefore it is feasible to arrest the progress of many explosions.

Suppressants may be in the form of a liquid, mist, or powder. The two most common types of suppressant are the halons, such as chlorobromomethane (halon 1011) and ammonium phosphate powder, but water is also used in some circumstances. The action of a suppressant is complex; several effects can be identified, any one of which could be responsible for the suppression of a particular explosion. The effects include (i) cooling of the combustion zone, for example, by evaporation of the liquid suppressant; (ii) free radical scavenging—active species in the suppressant inhibiting the chain of reactions by which the combustion propogates; (iii) advanced inerting—the concentration of suppressant in the unburned mixture can render it nonflammable; (iv) oxygen exclusion; (v) physical quenching—unburned particles or droplets are caused to agglomerate such that nonexplosive conditions prevail.

Suppression may be used in the following circumstances:

(i) The measured dust explosibility gives a K_{st} value less than 200 bars. m/s.

(ii) The material is nonmetallic.

(iii) The system may have a pressure detector set at less than 0.1 bar (i.e., normal system pressures do not rise above this value).

(iv) The suppressors are arranged to give effective dispersion of suppressant within 100 ms.

(v) The plant is shock resistant to a pressure pulse of 1 bar. (This may be reduced to 0.25 bar in some cases by using lower detection pressures and very rapid dispersal of liquid suppressants).

Suppression tends to be used as an alternative to venting where the geometry of the plant makes effective venting difficult. For example, a duct structure, such as in a flash drier cannot be vented effectively. Alternatively, if the plant is situated in a position where safe venting to the outside is impractical or where release of the product during venting would cause problems, suppression is often the best alternative.

Suppression is more expensive to install and maintain then venting but there is only minor contamination following an incident, and a major cleaning operation is not normally required. Suppression protects not only the plant itself, but also operators working in the area. There is no need to provide a safe area into which flames can vent; suppression can help avoid secondary explosions that tend to occur in rooms where dust deposits are inevitable.

7.4.2. The Suppression System

A complete suppression system for a plant has to be carefully and individually designed for each installation. It will consist of detectors in any region where an ignition may occur. These initiate the release of suppressant in the plant area where the ignition has been detected and in interconnecting ducting. They may also cause high-speed isolation valves to close and switch off the plant.

Detectors may operate either by a light source or by detecting the initial pressure rise. Although light detectors are commonly used in gaseous systems, pressure detectors are more usual in dust systems because the dust cloud may obscure a light source from the detector. Pressure detectors may be designed to detect either a steady rise in pressure or the situation where the rate of pressure rise exceeds a stated level. The latter type of detector is used in plants where normal pressure

FIGURE 7.8. Explosion pressure threshold detector. (Published with permission of Graviner, Ltd).

fluctuations occur that could exceed the threshold level for activation of the standard pressure detector. An explosion pressure threshold detector is shown in Figure 7.8.

Typical settings for the threshold and rate of rise detectors are 0.035 bar and 0.8 bar/s, respectively. The settings for activation of a suppression system are required to be as low as possible in order to minimize the overpressure, if ignition occurs. However, it is also important that normal fluctuations in pressure should not activate suppression accidently. This problem may be overcome by having two detectors mounted perpendicularly. The suppressant is activated only in a coincident signal from both detectors.

The suppressant may be contained in hardware of a number of different designs. Two examples are shown in Figure 7.9. Figure 7.9(a) shows a sectioned hemispherical supressor for a liquid suppressant. It can operate extremely rapidly, delivering the suppressant within 5 ms of detection with an initial discharge velocity in excess of 200 m/s. The hydraulic shock is generated by an electrically fired explosive charge in the center of the suppressant. This type of installation is fitted inside the plant and is not suited to high-temperature processes. It is exceptionally fast but of limited range and is therefore best suited to small installations.

Figure 7.9(b) shows a high-rate discharge suppressor that can deliver either liquid or powdered suppressant. The suppressant is contained in a cannister under high pressure (20–120 bars) and released when a

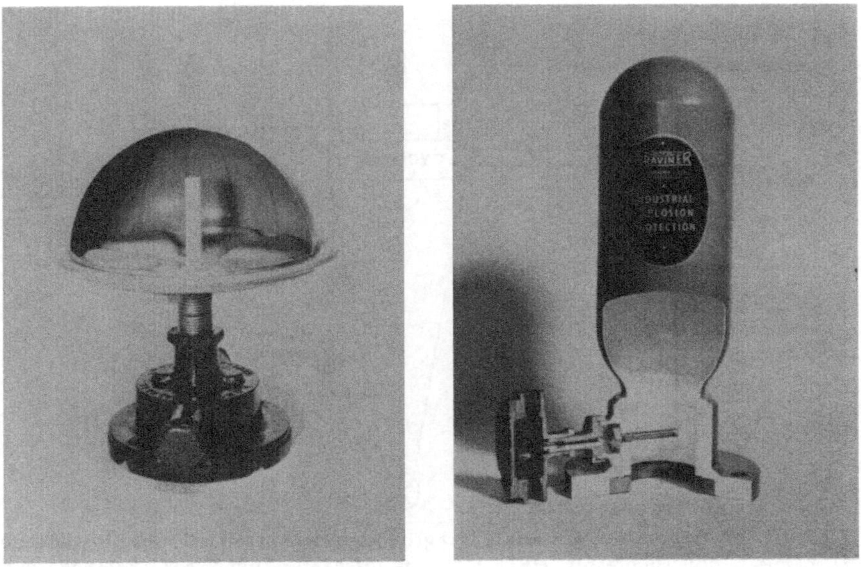

FIGURE 7.9. (a) Sectioned hemispherical suppressor for liquid suppressants. (b) Sectioned high-rate discharge suppressor for liquid suppressants. (Published with permission of Graviner, Ltd.)

frangible disk across the outlet is ruptured by a shock generated by an electrically fired detonator. The suppressant is ejected from the canister by compressed nitrogen. The delay in this type of device is typically 10–15 ms; it is used in larger installations where more time is available for suppression because of the lower rate of pressure development but where a greater suppressant throw is required.

Figure 7.10 shows typical explosion suppression counter measures for a grinder–cyclone system. Grinders are typical of applications where suppression is required because it is particularly difficult to infallibly prevent powder from overheating or eliminate tramp-metal sparks, and the design and positioning of the plant often makes it unsuitable for venting. In the example shown, the grinder and ducting are protected by hemispherical suppressors. Detector D_1 detects an explosion in the grinder and initiates the hemispherical suppressors. An explosion in the cyclone is detected by D_2 and suppressed by high-rate discharge suppressors. Flame propogation back to the grinder is prevented by activation of the duct suppressors and a high-speed isolation valve isolates this system from the rest of the plant. An additional suppressor is placed after the isolation valve to prevent flame propogation beyond the valve as it closes. In this case operation of either of the detectors is set to activate the whole system and also to shut down the plant.

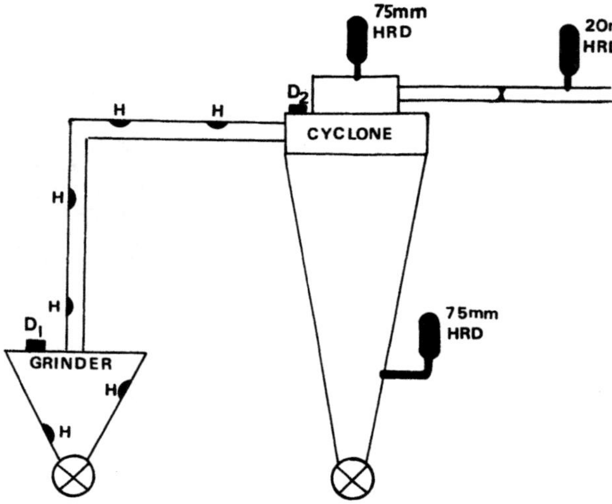

FIGURE 7.10. Suppression protection for a grinder–cyclone system. D, explosion detector, H, hemispherical suppressor, HRD, high-rate discharger suppressor. (Published with permission of Graviner, Ltd.)

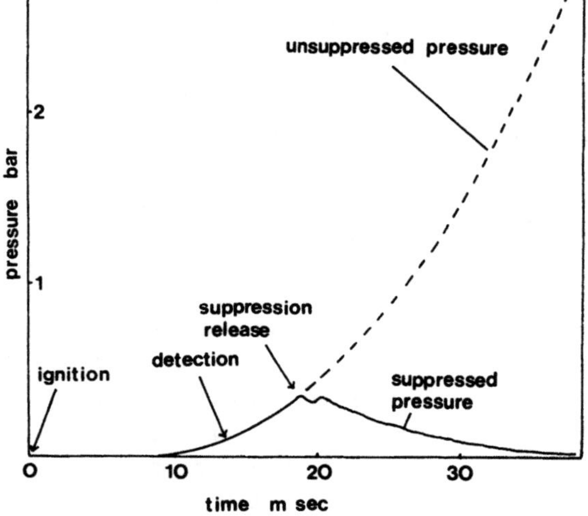

FIGURE 7.11. Development of suppressed and unsuppressed explosion.

The duration from ignition to suppression is typically 300 ms. Figure 7.11 shows a comparison of the development of the pressure in a suppressed and unsuppressed system.

7.4.3. Comparison of Suppressants

Halogenated hydrocarbons—known as halons—can be used as inerting agents to prevent explosions and also as suppressants. The different types of halon are given numbers that designate the relative number of carbon, fluorine, chlorine, and bromine atoms. Thus halon 1011 has 1 carbon atom, no fluorine atoms, 2 chlorine atoms, and 1 bromine atom. The halon is stored in liquid form and rapidly vaporizes when ejected.

Water can be an effective suppression system for dust explosions. Tests in a 1-m^3 sphere reduced the maximum pressure by an order-of-magnitude to between 0.6 and 0.9 bar. The third class of suppressants are powder systems, the most effective of which are based on ammonium phosphate or sodium bicarbonate powders.

The effectiveness of the three most commonly used explosion suppressants has recently be evaluated by British, American, German, Swiss, and French members of a working group of the International Standards Organization (Moore, 1979). The following observations were made:

(i) Water is not an effective suppressant of gas explosions but is effective against some dust explosions.

(ii) A 75-mm high-rate discharge system is more effective than a double-outlet 20-mm high-rate discharge system.

(iii) More effective suppression was achieved in larger tests volumes where more time was available to suppress the explosion.

(iv) In general, halon 1011 and ammonium phosphate powder (tropolar) had similar levels of effectiveness. However, some tests showed that halon 1011 could fail to suppress if a high detection pressure was used.

Figure 7.12 shows the maximum overpressure in a cellulose explosion in a 1-m^3 chamber using the three different suppressants as a function of the detection pressure. This clearly shows that a low detection pressure must be set for a halon suppressant. If it is released late, suppression may fail to occur. Figure 7.13 compares the halon initiated at 0.005 MPa and the ammonium phosphate initiated at 0.01 MPa as a function of the concentration of suppressant.

In general, a powder system is always initiated at a higher pressure. This reduces the incidence of plant shutdown from false alarms but does

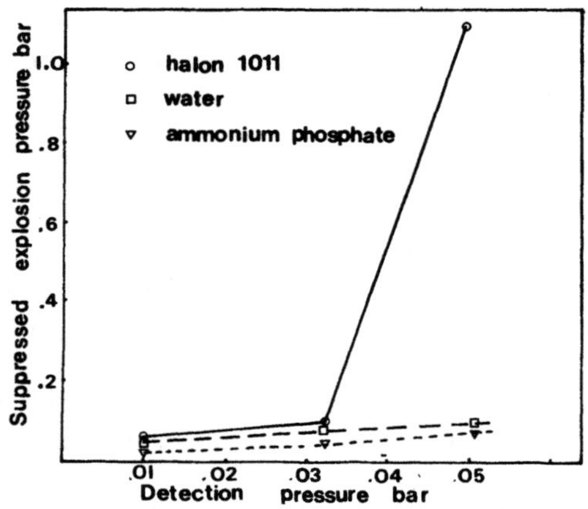

FIGURE 7.12. Suppressed pressure as a function of detection pressure for three different suppressants.

cause a larger overpressure so that the plant must be built stronger. The choice between liquid and powder suppressant is often fixed by the practicality of allowing the slightly higher overpressure against the nuisance and expense of initiation by a false alarm.

A liquid suppressant can cause a vacuum to be created if used in conjunction with an automatic door since air may be replaced by the suppressant gas which condenses on cooling. There is also some evidence that powder suppressants are more effective in large chambers (60 m^3) (Bartknecht, 1981). The work investigating the effectiveness of different suppression systems in chambers of different volumes has allowed a

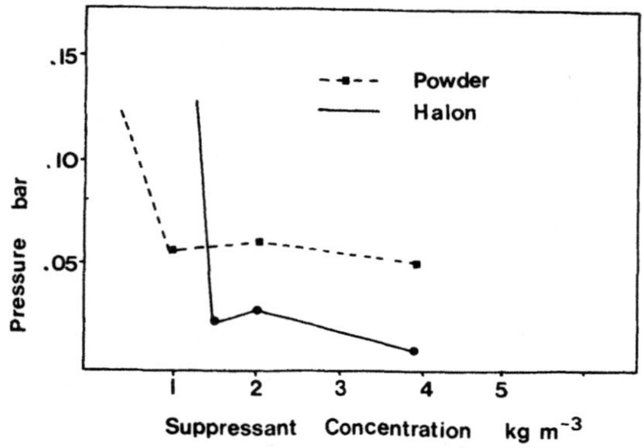

FIGURE 7.13. Pressure produced in 1-m^3 sphere in a suppressed cellulose dust explosion. Ammonium phosphate powder activation pressure, 0.1 bar; suppressant pressure, 60 bars. Halon 1011 activation pressure, 0.05 bar, suppressant pressure 20 bars.

mathematical model to be developed. This can be used to calculate the suppressant requirement characteristic as a function of time necessary to ensure effective suppression in a "worst case explosion." The model may be used to assist in the design of an industrial suppression system. Normally the industrial system is installed to give twice the predicted required delivery rate and a total mass of suppressant which is 50% greater than that required by the model.

7.5. INERTING

An inerting system reduces the oxygen concentration in the plant to a level that cannot support the explosion. The most common gases used are nitrogen and carbon dioxide, chosen for cheapness and availability but in some manufacturing processes a supply of inert gas from combustion products may be readily available, and this can provide a more economical solution to inerting in these cases. Inert gas generators can also be purchased. These generally burn oil to produce a low-oxygen gas which must be cooled and cleaned before use.

Generally, inerting is the most expensive solution to the dust explosion problem, but where ignitions are frequent and cannot be reduced it may be more economical in the long run to eliminate explosions completely by inerting rather than have the expense of frequent activation of a suppression system or the disruption and loss of time and product in a vented explosion.

An inerted system is usually a closed-loop design in which the inerted air output from the system is passed through a filter and recirculated Reduction of the oxygen concentration to a sufficiently low level is likely to involve replacing the volume air three times, and it can be seen that the consumption of inerting gas on a system which is not a closed-loop would be considerable.

In any system there will be losses and it may be necessary to introduce the inerting gas at more than one point in the plant. Losses can be minimized by ensuring that the pressure is as close to atmospheric as possible at points where complete sealing of the plant may be suspect.

Effective inerting relies on a low oxygen concentration throughout the plant. The oxygen concentration must be monitored continuously at a number of points throughout the plant and the detection system should be failsafe—to ensure that there is no hazard in the event of an oxygen monitor failure.

In a powder system, the powder itself contains about 50% by volume of air which will considerably dilute the inerting gas as the powder is fed into the system. An inerted entry system or a slow feed into a system at very low oxygen concentration is often necessary.

The level to which the oxygen content must be reduced is defined by tests for maximum permissible oxygen concentration described in Chapter 5. The maximum concentration is a function of the nature and size of the dust, the temperature of the plant, and the inerting gas being used. It is also a function of the ignition source and more oxygen can be allowed for a spark source than for a general heat source. It is therefore necessary to ensure that the detailed circumstances of the test for maximum oxygen concentration are known before using published data for defining the acceptable concentration and quantity of inerting gas required.

Early tests, for example, used a furnace apparatus and found that oxygen concentrations in carbon dioxide as low as 7–10% were required. Later, room-temperature tests gave figures of 10–13% in nitrogen and 13–16% in carbon dioxide.

NFPA regulation 69 suggests that maximum oxygen concentration in carbon dioxide using ignition with a spark source should be taken as the norm. Figures for an extensive range of dust are published by the United States Bureau of Mines. The maximum oxygen concentration should be reduced by 2% if the likely ignition source is of higher energy than a spark, such as an electrical arc. A hot surface ignition source should have a 6% safety margin allowed for high-temperature applications, or when the ignition source could be an open fire, a flame, or a glowing furnace wall. Temperature is an extremely important parameter and even small temperature rises of 100°C can significantly affect the quantity of inerting gas required.

As the quantity of oxygen is reduced, the maximum explosive pressure and the rate of pressure rise registered in an explosion decrease almost linearly. The minimum explosive concentration of dust is barely affected but the maximum explosive concentration decreases considerably so the band of explosive conditions becomes much narrower as the oxygen present decreases. The requirement of inerting gas will be affected by the maximum withdrawal rate, temperature changes, leaks (both the escape of inerting gas and the ingress of air), and rapid atmospheric changes.

Nitrogen is probably the most commonly used inerting gas, but carbon dioxide is slightly more effective and will allow a few percent more oxygen to be present (Table 7.3). A rule-of-thumb formula for calculation of the maximum permissible oxygen concentration when nitrogen is used instead of carbon dioxide as a diluent is given by

$$O_n = 1.3O_c - 6.3$$

TABLE 7.3. Maximum Permissible Oxygen Concentration with Nitrogen and Carbon Dioxide as Inerting Gases.

	Maximum oxygen concentration (%) (at room temperature)	
Material	CO_2 spark	Nitrogen (pyrotechnic igniter)
Coal	16	14
Wood	16	11
Methyl cellulose	13	10
Resin	14	10
Aluminum	2	4–6

where O_n is the maximum permissible oxygen concentration with nitrogen and O_c is the equivalent figure for carbon dioxide.

Carbon dioxide is heavier than air, therefore the plant can be filled by displacement more easily. However, carbon dioxide can have corrosive properties and is often less easily obtainable in the quantities required.

It is possible to use active suppressants such as the halons as inerting agents; these tend to be relatively more effective as they have a positive inhibiting effect rather than simply preventing access of oxygen. However, they are also far more expensive. It should be noted that some halogenated hydrocarbons and carbon dioxide can react with some dusts such as highly reactive metals or be ineffective at providing inertion. Argon and helium are thus used for inerting in special circumstances.

7.6. FLAME TRAPS AND AUTOMATIC BARRIERS

If an explosion or fire occurs in a flammable gas the flame is unable to pass through a small orifice. This forms the basis of flame traps or barriers. Flameproof electrical equipment is built so that all gaps are such that a flame inside the equipment cannot propogate to a region of flammable vapor outside. Generally the allowable gap is of the order of 1 mm.

Flame traps in gas lines may consist of a honeycomb structure of metal foil or a wire wool design. This type of flame barrier cannot be used directly for flammable dusts. Obviously a conventional flame trap in a powder pipeline would inhibit flow, but even in air-return lines in powder systems a flame trap can be ineffective because, although the flame is stopped, particles of burning powder can be blown through. For the same reason flameproof equipment is not adequate protection in a dusty environment.

For powder pipelines an extinguishing barrier may be used. The flame front can be detected and an extinguishing medium released. The release of the extinguisher may be triggered by either pressure detection or an optical signal. In pipes an optical system is often preferred because there will not be problems detecting light through a dust cloud as in large vessels and the relationship between the pressure front and the flame front in pipes is not sufficiently clear for accurate design of the delay time required. The optical sensor may detect either visible light or infrared radiation.

The detection system may also be used to activate rapid-action valves rather than to release extinguisher. Mechanical devices of this type are effective but care must be exercised to ensure that their action is not inhibited by contamination when they are used in dust-carrying lines. Slide vales in which the moving parts are located outside the cross section of the line may be more suitable in this application. Most automatic valves are activated by air released on activation of a small detonator triggered by a signal from the detection system. However, barrier valves have also been designed in which the pressure wave of the explosion pushes the valve closed. A certain minimum pressure is required to operate these valves and if the maximum overpressure (for example, in a vented system) is not likely to exceed the activation pressure then a detonation type of valve must be used.

Automatic valves are useful not only in preventing the spread of explosions but in sealing off systems in the event of a fire. For example, booth fires are not uncommon in electrostatic powder-coating systems but they are self-extinguishing because the dust concentration is suitable over only a small volume of the booth. However, if burning particles can get into the recirculatory system where there can be a confined dust cloud a serious explosion can occur. Automatic valves triggered by detection of a fire in the booth can prevent this occurrence. The optical detector must be kept clear of powder layers in order to operate effectively.

Appendixes

A1. SUMMARY OF ELECTROSTATIC EQUATIONS

Basic Relationships Between Charge, Field, and Voltages

$E = \dfrac{q}{4\pi\epsilon_0 r^2}$	Coulomb's Law—field due to a point charge
$E = \dfrac{Q}{\epsilon_0}$	field due to a surface charge Q per unit area
$E = -\dfrac{dV}{dx}$	$E = -\operatorname{grad} V$
$V = \dfrac{Q}{C}$	
$\nabla^2 V = \dfrac{Q}{\epsilon\epsilon_0}$	Poisson's equation
$\int E\, dS = \dfrac{\Sigma q}{\epsilon_0}$	Gauss's theorem
$q = \int I\, dt$	or for constant current, $q = It$ (by definition)

Resistivity and Resistance

$R = R_1 + R_2 + R_3$	resistances in series
$\dfrac{1}{R} = \dfrac{1}{R_1} + \dfrac{1}{R_2} + \dfrac{1}{R_3}$	resistances in parallel
$\rho = \dfrac{1}{\sigma}$	
$\rho = \dfrac{RA}{d}$	
$\tau = \epsilon\epsilon_0 \rho = RC$	time constant for charge decay
$R = \dfrac{V}{I}$	definition of resistance

Capacitance

$C = \frac{\epsilon\epsilon_0 A}{d}$	two planes
$C = 4\pi\epsilon\epsilon_0 a$	isolated sphere of radius a
$W = \frac{1}{2}CV^2 = \frac{1}{2}QV = \frac{1}{2}\frac{Q^2}{C}$	energy stored in a capacitance
$\frac{1}{C} = \frac{1}{C_1} + \frac{1}{C_2} + \frac{1}{C_3}$	capacitances in series
$C = C_1 + C_2 + C_3$	capacitances in parallel

A2. CERTIFICATION AND STANDARDS

Over recent years each of the developed countries have built up standards, codes of practice, and certification systems applicable to hazardous locations. Not all agree in detail and terminology is confused. Very recently some attempt has been made to coordinate standards internationally. However, it is proving to be difficult and time consuming even to produce consensus codes within Europe. The result is that, at present, there are still many codes covering each topic.

Hazardous areas are classified into zones (in Europe) or divisions (in the United States). The British system has three zones:

(a) Zone 0 where a flammable atmosphere is continuously present;

(b) Zone 1 where a flammable atmosphere is likely in normal operation;

(c) Zone 2 where a flammable atmosphere is unlikely in normal operation and if it occurs will exist only for a short time.

The Americans use only two divisions—flammable atmosphere likely and unlikely to occur. There is no Zone 0. France and the Netherlands also use the two division systems known in France as Zone E and F. The nature of the hazardous atmosphere is also classified and again the terminology is not universal.

In the United States, Class I hazards are gases and vapors; Class II are dusts; and Class III are fibers. Class II is further subdivided into E, F, and G according to the electrical resistance of the dust.

In Britain, dusts are divided into groups A or B according to whether they ignite in specific test equipment (Chapter 5). This very simple classification aims simply to define whether or not protective measures must be taken in industry. If a dust is defined as ignitable then it must be further subdivided to define the degree of hazard in the event of ignition. The parameter chosen for this purpose is the rate of pressure rise in the event of ignition. The rate of pressure rise defines the overpressure that will occur in a vented vessel and also defines the severity of the explosion and the difficulty of designing suitable protective measures.

The German system of powder classification is now accepted throughout Europe except in the United Kingdom where it is gradually gaining acceptance. The United States has adopted a very similar system. In Germany, dusts are tested in a 1-m³ sphere. The rate of pressure rise measured in this vessel can be extrapolated to larger, similar-shaped volumes according to a Cubic Law

$$\frac{dP}{dt} V^{1/3} = K_{st}$$

The constant K_{st} is a constant of the powder sample. If K_{st} is less than 200 the dust is Class I, if K_{st} is between 201 and 300 it is Class II, and if above 300 the dust is Class III.

In the German system is it assumed that K_{st} does not depend on ignition source and this assumption appears reasonably justified provided a continuous arc is not used. The Americans have introduced a system more directly analogous to gases which is identical to the German system for high-energy ignition sources. However, with low-energy ignition sources, the classification system is defined as Class I for K_{st}-1–100, Class II for K_{st}-101–200, and Class III for K_{st} above 200.

Standards and codes of practice for dealing with processes involving flammable dusts also differ from country to country. The organizations for each country which publish Codes of Practice and standards are given in Table A1 and a list of relevant codes in Table A2.

TABLE A1. Organizations Issuing Codes of Practice and Standards

Country	Name and address of organization
United States	National Fire Protection Association 470 Atlantic Avenue Boston, Massachusetts 02210
	American National Standards Institute 1430 Broadway New York, New York 10018
	Underwriters Laboratory 1285 Walt Whitman Road Melville, L.I., New York 11746
Canada	Standards Council Canada International Standards Branch 2000 Argentia Road Suite 2/ 401 Mississauga, Ontario 15N 1V8

(continued)

TABLE A1. (*cont.*)

Country	Name and address of organization
West Germany	P.T.B. (Physikalische Technische Bundesanstalt) 33 Braunschweig Bundesalle 100
	D.I.N. (Deutches Institut für Normung) Beuth Verlag BmbH Postfach 1107 Burggrafenstrasse 4/10 1000 Berlin 300
France	A.F.N.O.R. Tour Europe Cedex 7 Paris le Defense
United Kingdom	British Standards Institute British Standards House 2 Park Street London W1
	Safety in Mines Research Establishment Health and Safety Executive Redhill Sheffield S3 7HQ
Europe (general)	CENELEC (European commmittee for electrotechnical-standardization) Rue Brederode 2 Bte 5 1000 Brussels Belgium
	I.E.C. (International Electrotechnical Commission) 1 Rue Varembe Geneva Switzerland
Australia	Standards Association of Australia Standards House 80 Arthur North Sydney New South Wales

TABLE A2. Standards and Codes of Practice Relating to Fires and Explosions in Powder Systems.

	British Standards
B.S. 3013	Fire precautions in chemical plant
B.S. 138, 336, 740, 750, 1382, 1721, 3326, 3465, 3709, 5306, 402	Fire extinguishers
B.S. 1259	Intrinsically safe electrical apparatus and circuits for use in explosive atmospheres (Amendments 1961)
B.S. 1821	Welding of steel pipelines carrying fluids.
B.S. 263	Arc welding of ferite steel pipework carrying fluids.
B.S. 2050	Electrical resistance of conducting and antistatic products made of flexible polymeric material.
B.S. 3187	Electrically conducting rubber flooring.
B.S. 5951	Electrically conducting and antistatic footwear.
CP 1013	Grounding
B.S. 4137	Guide to the selection of electrical equipment for use in Division 2 areas.
B.S. 4547 (E.N.2)	Classification of fires.
B.S. 4683	Electrical apparatus for explosive atmospheres.
B.S. 5000	Rotating electrical machines of particular types for particular application. Part 16 Type N electrical motoring.
B.S. 5345	Selection installation and maintenance of electrical apparatus for use in potentially explosive atmosphere.
B.S. 5545	Automatic fire detection systems.
B.S. 5458	Safety requirements for indicating and recording electrical measuring instruments and their accessories.
B.S. 5490	Protection for enclosures.
B.S. 5501	Electrical apparatus for potentially explosive atmospheres.
B.S. 5839 Pt. 1	Fire detection and alarm systems in buildings.
CP 326	The protection of structures against lightning.
CP 1003	Electrical apparatus and associated equipment for use in explosive atmospheres of gas or vapor (other than mining applications).
	European
CENELEC	EN 50 020 (equivalent to B.S. 5501 pt 7).
I.E.C. 144	Degression of protection of enclosures for low-voltage switch gear and control gear.
I.E.C. 31	Electrical equipment for explosive atmospheres.
I.E.C. 79	Electrical apparatus for explosive gas atmospheres.
I.E.C. 93	Recommended method of test for volume and surface resistivities of insulating materials.
D.I.N. 54	Testing of textiles—electrostatic assessment.
	Safety in Mines Research Establishment, Sheffield
Digests	
1. Testing of intrinsically safe apparatus.	

(continued)

TABLE A2. (*cont.*)

3. On load testing of flameproof electrical equipment.
6. High voltage equipment for use in flammable atmospheres.
7. Testing of flameproof electrical equipment.

Safety in Mines Research Paper, 104, 1946.
Intrinsic safety of electrical apparatus H.M.S.O. (U.K.) 10.

BASEEFA Standard

Intrinsic Safety Standard S.F.A. 3012.
"Certification standard for intrinsically safe apparatus."

N.F.P.A. (U.S.A.)

497	Hazardous locations for electrical installations in chemical plants classification of:
77	Static electricity—recommended practice
68	Guide to explosion venting
10–16	Fire extinguishers
493	Intrinsically safe apparatus
496	Purged enclosures for electrical equipment
51	Cutting and welding
61A	Manufacturing and handling starch
61B,C,D	Grain elevators, feed mills—agricultural commodities
63	Industrial plant—dust explosions
651	Aluminum or magnesium powder
653	Coal preparation plants—dust hazards
655	Sulfur fires—explosion prevention
656	Spice grinding plants—dust hazards
66	Pneumatic conveying systems
69	Explosion prevention systems
72E	Automatic fire detection
704	The fire hazards of materials
1	Fire prevention code
325	Properties of flammable liquids, gases, and solids
49	Hazardous chemicals data
491	Hazardous chemical reactions
	Instrument Society of America, Pittsburgh, ISA S.P 12 10, Area classification and hazardous dust location.
	Underwriters Laboratories; American National Standards A.N.S.I. C33 93, 1977; Standard for electrical motors and generation for use in hazardous location class II groups E, F, G.

A3. SOURCES OF EXPLOSION DATA

United States

U.S. Bureau of Mines—Hartman Bomb Tests

Minimum ignition energy, maximum explosion pressure, maximum rate of pressure rise, maximum oxygen concentration, minimum explosive concentration; Ignition temperature of a layer, ignition temperature of a cloud (furnace apparatus).

U.S. BUREAU OF MINES REPORTS

4725		Explosibility of cornstarch
5624	1960	Laboratory equipment and test procedures
5754	1961	Explosibility of agricultural dusts
5971	1961	Explosibility of dusts used in the plastics industry
6516	1964	Explosibility of metal dusts
6597	1965	Explosibility of carbonaceous dusts
7132	1968	Explosibility of chemicals, drugs, dyes and pesticides
7208	1968	Explosibility of miscellaneous dusts

National Fire Protection Association

325. Properties of flammable liquids, solids, and vapors. Flash point, ignition temperature, flammability limits, extinguishing method.

68. Explosion venting guide. Minimum ignition energy, minimum explosive concentration, maximum explosive pressure, and rate of pressure rise measured in Hartman apparatus.

69. Explosion prevention systems. Maximum allowable oxygen and concentration ignited by a spark with carbon bioxide as atmospheric diluent.

Factory Insurance Association, Chicago

Booklet on Dust Explosions, Analysis, and Control. Ignition temperature of a cloud, minimum ignition energy, minimum explosive concentration, maximum explosive pressure, maximum rate of pressure maximum permissible oxygen concentration. (excerpts from U.S. Bureau of Mines Report for more common materials).

Europe

Bren-und Explosions—Kenngrossen von Stauben, Bonn, W. Germany

Maximum pressure and rate of pressure rise, minimum explosive concentration, temperature of cloud, temperature of layer, maximum permissible oxygen concentration, measurements made on many size distributions using a 1-m^3 sphere.

K.N. Palmer—Dust Explosions and Fires

Chapman and Hall (1973). Minimum ignition energy, ignition temperature of cloud and layer, minimum explosive concentration, maximum pressure and rate of pressure rise. (Results of work at the Fire Research Station, U.K.)

Fire Research Station, U.K.

Dust Explosion in Factories—classified list of dusts that have been tested for explosibility. SHW 830 (Department of Employment, U.K.). Ignition energy, minimum concentration, maximum pressure and rate of pressure rise (tested in Hartman apparatus), ignition temperature of cloud and layer.

Sources of Hazards Data

L. Bretherick (1975). *Handbook of Reactive Chemical Hazards*. Butterworth, London.

N.I. Sax (1975) *Dangerous Properties of Industiral Materials*, 4th Ed. Von Nostrand Reinhold, London.

Washington National Library of Medicine Toxilino Chemical Dictionary.

CHETA Program. American Society for Testing and Materials. Philadelphia.

Dow *Safety and Loss Prevention Guide*, 3rd ed. 1973. Chemical Engineering Progress Tech Manual. American Institute of Chemical Engineers, New York (Dow Fire and Explosion Index).

Fire Protection Guide of Hazardous Materials (1979) N.F.P.A. Boston, MA.

Bibliography

B1. FIRE

Introduction to fire, Fire Protection Association, London.

Good practice in the fire protection of buildings, Fire Protection Association, London.

Automatic fire detection and alarm systems, Booklet No. 41, Fire Protection Association, London (1964).

Automatic fire alarm systems, *Fire Protection Assoc. J.*, No. 66, JR-25.

Management guide to fire protection, Industrial Fire Protection Association, London.

Some modern developments in fire detection and extinction, by M. J. O'Dogherty, *Q. J. Inst. Fire Eng.*

Behaviour of structural steel in fire, Symposium 2, Department of Environment, London.

Automatic fire detection systems, by Q. Fordham Copper, *Industrial Systems and Equipment*, pp. 20–25, June (1967).

Codes of safe practice in the petroleum industry, Institute of Petroleum, London.

Fire prevention procedures in the use of plastics packaging, British Plastics Federation, London (1979).

Recommendations for the protection of computer installations against fire, Fire Officers Committee, Fire Protection Association, London (1972).

Recommended good practice for protection of electronic data processing and computer controlled industrial processes, Factories Insurance Association, Hartford, Connecticut (1971).

Smoke and toxic gases from burning plastics, by G. W. V. Stark, Building Research Establishment, Note CD 5/74, H.M.S.O., London.

Thermal Degradation of Organic Polymers, by S. L. Madorsky, John Wiley and Sons, New York (1964).

Fire fighting in factories, Health and Safety at Work, Booklet No. 10, H.M.S.O., London.

Portable fire extinguishers, Fire safety data sheets FS 6001, 6002, 6003, Fire Protection Association, London.

Hotwork welding and cutting on plants containing flammable materials, H.M.S.O., London.

Modern fire protection for structural steel work, Publication No. PP2, British Constructional Steel Assocation, London (1973).

Guide to fire protection in the chemical industry, Chemical Industries Association, London (1968).

Safety and management: a guide to the chemical industry, Chemical Industries Association, London.

B2. DUST EXPLOSIONS—GENERAL

Dust Explosions, by K. N. Palmer, Chapman and Hall, London (1973).

Dust Explosions Cause and Cure, by W. Bartknecht, Springer Verlag, Berlin and New York (1980).

Dust Explosions and Fires: A Guide to Literature 1957–1977, by P. Cardillo and E. J. Anthony, Stazione Sperimentals per i combustibili, S. Donato Milanese, Italy (1979).

(Fire research abstracts and reviews) *Fire and explosion hazards in the pharmaceutical industry* (also the electronics, rubber, textile, aerospace, paper and board, and paint and ink industries), National Academy of Sciences, Washington, D.C.

U.K. Data Sheets 6015, 6021, 6022, 6023, and 6024, Fire Protection Association, London.

A bibliography of topics related to the study of grain dusts: fires and explosions, published by Energy and Mineral Resources Research Institute, Iowa State University, Ames, Iowa (1976).

Dusts explosions in factories, Health and Safety at Work, Booklet No. 22, H.M.S.O., London.

Flammable vapours, gases and dusts in industry, Ventilation T.I.S. 5006, Fire Protection Association, London.

Area classification in hazardous dust locations, Instrument Society of America, Pittsburgh.

The dust explosion hazard and its prevention, by E. M. Murphy and J. Nagy, Monograph 112, Instrument Society of America, pp. 74–88, Pittsburgh (1969).

VDE 0165/8.69 *Bestimmungen fur die Errichtung elektrischer Anlagen in explosionsgefahrdeten Betriebsstatten*, Verband Deutsher Electrotehniker, Berlin (1969).

Measurement comes to hazardous dust area classification, by T. W. Moodie, Paper No. 70-666, 25th Annual Conference, Instrument Society of America, Pittsburgh (1970).

Inhibition of coal-dust air flames, by J. Gruner, U.S. Bureau of Mines R.I. 7552, Pittsburgh (1971).

Proceedings of the International Symposium on Grain Dust Explosions, Grain Elevator and Processing Society (1977).

Dust explosions, by R. F. Schwab, *Chem. Process Eng.* (London), p. 164 (1964).

B3. INDUSTRIAL PROCESSES

User guide to fire and explosion hazards in the drying of particulate materials, Institute of Chemical Engineers, Rugby, U.K.

Fire and explosion hazards in spray driers, by N. Gibson and F. Schoffield, *Institute of Chemical Engineers (U.K.) Symposium, Series 49,* 53 (1977).

Nature and prevention of dust and gas explosions in containers and pipes, by W. Bartknecht, *Chem. Ing. Tech.* **47** (6), 236 (1975).

Explosions in dust collecting plants, by K. N. Palmer, *Chem. Eng.*, p. 295 (1975).

Dust explosions hazards in pneumatic transport, by K. N. Palmer, *Fire Prevent. Sci. Tech.* **11**,4 (1975).

Explosion protection of dust extraction systems, by K. N. Palmer, *Institute of Chemical Engineers Symposium, Series 39,* Rugby, U.K. (1974).

Safety precautions for the prevention of dust explosions and fires at sanding lines in the pressboard industry, by M. Vohringer, *Staub Reinhalt. Luft* **31** (3), 44 (1971).

Fire and explosion hazards, by J. Cross, in *Pneumatic Transport of Polymers,* G. Butters Ed., Applied Science Press, Amsterdam and New York (1981).

The fire hazards of belt conveyors, Factory Mutual Engineering Corporation., Record No. 1977, p. 18.

B4. PLANT LAYOUT

Fire precautions in chemical plants, CP 3013, British Standards Institute, London (1974).

Plant Layout, T. C. Mecklenburg Ed., Leonard Hill, London, (1973).

Recommended practices Nos. 520 and 521, American Petroleum Institute, New York (1960).

B5. HAZARD ANALYSIS

A guide to hazard analysis and operability studies, Chemical Industry Safety and Health Council, Chemical Industries Association, London (1977).

CEP. Technical Manual, by W. W. Russell, Loss Prevention 10, p. 80, AIChE, New York (1976).

Fault tree synthesis for chemical processes, by G. J. Power and F. C. Tompkins, *Am. Inst. Chem. Eng. J.* **20** (2), 376–387 (1974).

Safety in Process Plant Design, by G. L. Wells, John Wiley and Sons, New York (1980).

Handbook of System and Product Safety, by W. Harmer, Prentice Hall, Englewood Cliffs, New Jersey (1972).

Loss prevention and safety promotion in the process industries, by G. A. Martin Solis, P. K. Andow, and F. P. Lees, DECHEMA, Frankfurt (1977).

Numerical Analysis of hazards and failures, by P. K. Andow, F. P. Lees, and C. P. Murphy, *Int. Chem. Eng. Symposium (U.K.) Series 58*, p. 23, Rugby, U.K. (1980).

B6. TEST METHODS

Laboratory studies on spontaneous heating of coal, by A. G. Kerns, *U.S.* Bureau of Mines I.C. 8756, Pittsburgh (1977).

Temperature investigation of dust air mixtures to determine maximum safe temperature, by Slawomir Hulanickt, *Staub Reinhalt. Luft* **34** (11), 300 (1974).

Thermal Methods of Analysis, by W. W. Wendlondt, John Wiley and Sons, New York (1974).

Optical probe for measurement of dust concentration, by I. Liebman, *New Sci. Instrum.* **48** (10), 1314 (1977).

Tests for explosion resistance of electrical equipment, by U. Polka, *Electrotechnik* (Germany) 60 (7), 24–26 (1978).

Laboratory equipment and test procedures for evaluating the explosability of dusts, by H. G. Dorsett, U.S. Bureau of Mines R. I. 5624, Pillsbury, (1960).

Explosion tests for industrial dusts, by M. Raferty, Fire Research Technical Paper No. 121, H.M.S.O., London (1968).

Towards absolute minimum ignition energies for dust clouds, by. R. K. Eckhoff, *Combustion and Flame* **24**, 53 (1975).

Why are long sparks more effective dust explosion initiators than short sparks?, by R. K. Eckhoff, *Combustion and Flame* **27**, 129–131 (1976).

B7. WELDING

Electric arc welding, Health and Safety at Work, Booklet No. 38, H.M.S.O., London (1968).
Acetylene gas and oxygacetylene processes in factories, S.H.W. 1704, H.M.S.O., London
Electric arc welding, Department of Employment, Safety, Health, and Welfare, Series No. 38, H.M.S.O., London (1968).

B8. ELECTROSTATICS

Electrostatics and Its Applications, by A. D. Moore, John Wiley and Sons, New York (1976).
Electrostatics Hazards Evaluations and Controls, by H. Haase, Verlag Chemie, Weinheim and New York (1977).
Static Electricity, Industrial Safety Handbook, by N. Gibson, (W. Handley, Ed.), McGraw-Hill, pp. 107–121, London (1969).
Static electricity, *Safety in Industry Physical Hazards No. 8*. U.S. Department of Labor, Washington, D.C.
Electrostatics, by A. W. Bright, R. P. Corbett, and J. F. Hughes, Design Council, London (1978).
Sensitivity of explosives to initiation by electrostatic discharge, by F. W. Brown, D. J. Kusler, and F. C. Gibson, U.S. Department of Interior, Washington, D.C.
Proceedings of the Conferences on Static Electricity: Series 4, A. C. Strickland, Ed. (1967); Series 11, D. K. Davies, Ed., (1971); Series 27, A. Blythe, Ed. (1975); Series 48, J. Lowell, Ed. (1979); Institute of Physics, London.

B9. IMPACT SPARKS

Ignition of gases and vapors by friction and impact, by F. Powell, *Ind. Eng. Chem.* **61** (12), 29 (1969).
The use of light metals and their alloys in hazardous areas, E.R.A. Report 70-32, Electrical Research Association, Leatherhead, U.K.
Sparking hazards from aluminum paint, Safety Information Bulletin No. 9, Unilever Ltd.
The ignition of flammable gases by sparks from aluminum paint and rusty steel by F. E. T. Kingman, *J. Appl. Chem.* **2**, 449 (1952).
Aluminum paint hazards: the thermite type reaction, Nuclear Regulatory Commission, Washington, D.C.

B10. ELECTRICAL EQUIPMENT

Purged and pressurized containers for electrical equipment in hazardous locations, N.F.P.A. No. 496, National Fire Protection Association, Boston (1974).

National Electrical Code 502-503, National Fire Protection Association, Boston.

Enclosures for use in Class II groups E, F, G; Hazardous Locations, Canadian Standards Association C22.2 No. 25, Ottawa, Ontario (1966).

Industrial control equipment for use in hazardous locations, Underwriters Laboratory, Inc., Standards for Safety No. 698, Chicago.

Outlet boxes and fillings for use in hazardous locations, Standards for Safety No. 886, Underwriters Laboratory, Chicago.

British, European, and International standards reviewed, *Electrical Times* 13/20 April, No. 4518, pp. 8–9 (1979).

Guide to electrical installations in flammable atmospheres, Royal Society of Prevention of Accidents, Cannon House, The Priory, Queensway, Birmingham.

A user guide to intrinsic safety measurement technology, Application Note AN 9003, Luton, U.K. (1976).

A user guide to zone 2 instrumentation measurement technology, Application Note AN 9004, Luton, U.K. (1977).

Electrical instruments in hazardous locations, 3rd Ed., by Ernest C. Magison, Instrument Society of America, Pittsburgh (1978).

B11. THERMAL IGNITION

Applications of the theory of thermal explosion to the self-heating and ignition of organic materials, by P. C. Bowes, Fire Research Note 867, Fire Research Station, Boreham Wood, U.K. (1971).

On self-ignition of dust deposits, by T. Grewer, *Staub Reinhalt. Luft* **31** (3), 7 (1971).

Ignition of combustible dusts on hot surfaces, by P. C. Bowes and S. E. Townsend, *Br. J. App. Phys.* **13**, 105 (1962).

Thermal Degradation of Organic Polymers, by S. L. Madorsky, John Wiley and Sons, New York (1964).

Thermal phenomena during ignition of a heated dust dispersion, by J. Nagy and D. Surincik, U.S. Bureau of Mines R.I. 6811, Pittsburgh (1966).

P. C. Bowes, in *Combustion and Flame* **12**, 289 (1968).

Laboratory studies on self-heating of coal—a summary of information in the literature, U.S. Bureau of Mines I.C. 8757, Pittsburgh (1977).

The spontaneous ignition and dust explosion hazards of certain soyabean products, Bulletin No. 47, Underwriters Laboratories, Chicago.

B12. VENTING OF PLANT

Dust explosions, prevention and control, by R. E. Cocks, *Chem. Eng.* **86** (24), 94 (1979).

Explosion venting for low pressure equipment, by E. A. Rust, *Chem. Eng.* **86**, 102, November (1979).

Release of the pressure of an explosion with rupture disks and explosion valves, by C. Donat, paper presented at *A. Chema.*, Frankfurt.

Application of explosion pressure relief as a protective measure for industrial plant equipment, by C. Donat, *Loss Prev.* **11**, 87 II, AIChE, New York (1977).

Explosion protection measures on fluidizied bed spray granulators and fluidized bed driers, Central Safety Service, Ciba Geigy Ltd., Basel, Switzerland (1975).

Calculating a pressure–time diagram for an explosion vented space, by W. G. Chappell, Loss Prevention II, AIChE, New York (1977).

Dust Explosions in factories: the protection of plant of hinged explosion doors, by K. C. Brown and D. G. Wilde, Research Report 119, Safety in Mines Research Establishment, Buxton, U.K. (1955).

Explosion pressure relief, by W. Bartknecht, *Loss Prev. II*, P. 93, AIChE, New York (1977).

Pressure relief as used in explosion protection, by C. Donat, *Loss Prev. II*, P. 87, AIChE, New York (1977).

Explosions—Cause, Prevention, Protection, by W. Bartknecht, (Translated by Burg and Almond), Springer Verlag, Berlin (1981).

Dust Explosions, by K. N. Palmer, Chapman and Hall, London (1973).

Explosion venting guide, N.F.P.A. 68, National Fire Protection Association, Boston (1978).

Efficacite des events d'explosion, Note No. 881-74-74, by J. Pineau, *Cahiers de Notes Documentaires* No. 74, er trimestre, Institute Nationale de Rechereché et de Securité, Paris (1974).

Dust explosion venting—a reassessment of data, by K. N. Palmer, Fire Research Note 830, Fire Research Station, Boreham Wood, U.K. (1970).

Explosion protection of a dust extraction system, by K. N. Palmer, *Institute of Chemical Engineers Symposium, Series 39*, Rugby, U.K. April (1974).

Explosion development in closed vessels, by J. Nagy, E. C. Seigler, J. W.

Conn, and H. C. Verakis U.S. Bureau of Mines R.I. 7507. Pittsburgh (1971).

B13. VENTING OF BUILDINGS

Explosible dusts: control of explosions, Fire Safety Data Sheets FS 6022, Fire Protection Association, London, p. 5 (1971).

Explosion venting guide, by D. A. Decker, *Fire Tech.*, pp. 219–223 August (1971).

Explosion venting of low-strength equipment and structures, by C. Yao, *Loss Prev.* **8**, AIChE, New York (1974).

Flammable liquids and gases: ventilation, Fire Safety Data Sheets FS 6013, Fire Protection Association, p. 5, London (1970).

Reducing explosion damage by explosion venting and explosive protection systems, Special Interest Bulletin No. 156, American Insurance Association, p. 2, New York (1974).

The relief venting of dust explosions in process plants, by K. N. Palmer, Fire Research Note 1971, Fire Research Station, Boreham Wood, U.K.

Design of an explosion relief system for a building handling liquified fuel gases, by D. J. Rashbash, D. D. Drysdale, and N. Kemp, *Proceedings of a Symposium Organized by the Institute of Chemical Engineers, Scottish Branch, at the Howden Park Conference Center, September 14–15*, pp. 145–156, Institute of Chemical Engineers, Rugby, U.K. (1976).

Dust explosion venting: considerations of further data, by K. N. Palmer, Fire Research Note No. 961, p 16, Fire Research Station, Boreham Wood, U.K. March (1973).

A procedure for designing gas combustion venting systems, by W. B. Howard and W. W. Russell, *Symposium Series No. 39a: Proceedings of Symposium on Chemical Process Hazards with Special Reference to Plant Design V, North Western Branch of the Institute of Chemical Engineers at the University of Manchester, Institute of Science and Technology, April, 9–11*, p. 179, Rugby, U.K. (1974).

Recent advances in the relief venting of dust explosions, by K. N. Palmer, *J. Hazardous Matter.* pp. 97–111 January, (1976).

Venting explosions: an empirical approach, by M. A. Nettleton, *Fire Prevent. Sci. Tech.* March, 14, 27–31 (1976).

Bemessung von druckentlastongosoffnungen zum schultz explosionrefahdeter anbagen in der chemischen industrie, *J. Chem. Ing. Tech.* **38**, 1125 (1966).

G. Mundy, in *2nd Symposium on Chemical Process Hazards*, Institute of Chemical Engineers 46.

Sizing vents for dust explosions, by J. Singh, *Chem. Eng.* **86,** 103 (1979).
S. Crescitelli, G. Russo, and V. Tufano, in *J. Occupational Accidents* **2,** 125–133 (1974).

B14. SUPPRESSION

Characterization of dust explosability—a comparative study of test methods, by P. E. Moore, *Chem. Ind.*, p. 430 London (1979).
Dust explosion hazard assessment, by P. E. Moore, *Chemical Engineering Symposium (U.K.) Series 58, Chemical Process Hazards VII,* Institute of Chemical Engineers, p. 77, Rugby, U.K. (1980).
Institute of Chemical Engineering (U.K.) Series 58, Chemical Process Hazards VII, Institute of Chemical Engineers, p 171, Rugby, U.K. (1980).
Detection of explosions, by B. Ray, *Phys. Technol.* **9,** 235 (1978).
Suppression of dust explosions in containers, by W. Bartknecht, *Staub Reinhalt. Luft* **31** (3), 28 (1971).
Explosion suppression in industry, by P. E. Moore, *Phys. Technol.* **10,** 202 (1979).
Recent research concerning extinguishing coal dust explosions, by J. Gruner, *15th International Symposium on Combustion,* Combustion Institute, p. 103 Pittsburgh (1974).

B15. INERTING

A. Riurnara and P. Cardillo, in *Rev. Combustibili.*, p. 115 (1977).
Thermal phenomena during the ignition of heated dust dispersions, by J. Nagy, and D. J. Surincik, U.S. Bureau of Mines R.I. 6811, Pittsburgh (1966).
Preventing ignition of dust dispersions, by J. Nagy, H. G. Dorsett, and M. Jacobson, U.S. Bureau of Mines R.I. 6543 (1964).
Prevention of dust explosions by inerting, by J. Nagy, H. H. Dorsett, and M. Jacobson, U.S. Bureau of Mines R.I. 6561, Pittsburgh.
The toxicity of halon extinguishing agents: a new appraisal, by P. F. Thorne, Department of the Environment, Fire Research Note 1073, H.M.S.O., London (1977).

B16. CASE HISTORIES

Case histories of accidents, N.F.P.A. 60, National Fire Protection Association, Boston.
Case histories of accidents in the chemical industry, Manufacturing Chemists Association.

References

Chapter 1

Bartknecht, W., *Explosions Cause and Cure*, Springer Verlag, Weinheim and New York, (1980).

Chiotti, P., Literature survey of dust explosions in grain handling facilities: cause and prevention, Energy and Mineral Institute, Iowa State University Project 400-25-04, Iowa State University, Ames, Iowa (1976).

Jacobson, M., Nagy, Y., and Cooper, A. R., U.S. Bureau of Mines R.I. 5971, Pittsburgh (1962).

Chapter 2

Widgington, D. W., Some aspects of the design of intrinsically safe circuits, Research Report No. 256, Safety in Mines Research Establishment, Sheffield (1968).

Archard, T. F., The temperature of rubbing surfaces, *Wear* **2**,438 (1952).

Blinkensderfer, A., Methane ignition by friction and impact heating, *Combustion and Flame* **25**, 143–152 (1975).

Block, X., *General Discussions of Lubrication, Vol. 2,* Institute of Mechanical Engineers, London (1937).

Bowden, F. P., Initiation of solid explosives by friction and impact, *Proc. R. Soc. Lond.* **A198**, 337–349 (1949).

Bowden, F. P., Recent studies in metallic friction, Forty-First Thomas Hawksley Lecture, Institute of Mechanical Engineers, London (1954).

Bowden, F. P. and Thomas, P. H., The surface temperature of sliding, *Proc. R. Soc. Lond.* **A223**, 29 (1954).

Bowden, F. P., Stoke, M. A., and Tudor, G. K., Hot spots on rubbing surfaces and the detonation of explosives by friction, *Proc. R. Soc. Lond.* **A188**, 329 (1947).

Bowes, P. C., Thermal explosion of benzoyl peroxide, *Combustion and Flame* **12**, 289 (1968).

Bowes, P. C. and Townshend, S. E., Ignition of combustible dusts on hot surfaces *Br. J. App. Phys.* **13**, 105 (1962).

Brown, H. R., Dust explosion hazards in plants producing aluminum, magnesium, and zinc powders, U.S. Bureau of Mines I.C. 7148, Pittsburgh (1941).

Chaudhri, M. M., Stab ignition of explosives, *Nature* **263**, 121 (1976).

Field, J. E., The role of localized plastic flow impact initiation of explosives, *Proc. R. Soc. Lond.* **A343**, 399–413 (1975).

Gibson, N., Lloyd, F. C., and Perry, G. R., *Institute of Chemical Engineering Symposium (U.K.), Series 25* (1968).

Grice, C.S.W., Research paper, Safety in Mines Research Board 50/1952, U.S. Bureau of Mines, Pittsburgh (1952).

Heavens, S. N. and Fields, J. E., Ignition of a thin layer of explosives by impact, *Proc. R. Soc. Lond.* **A338**, 77–93 (1974).

Heighway, R. J. and Taylor, D. S., Transient temperature rises during the rubbing of metals on glass, *Wear* **9**, 310 (1958).

Holm, B., Calculation of the temperature development in a contact heated in the contact surface, and application to the problem of the temperature rise in a sliding contact, *J. Appl. Phys.* **19**, 361 (1948).

Holm, R., Temperature development in a heated contact with application to sliding contacts, *J. Appl. Mech* **19**, 369 (1952).

Jaeger, J. C., Moving sources of heat and the temperature at sliding contacts, *Proc. R. Soc. New South Wales* **76**, 203 (1942).

Kelley, J. S. and Forkner, B. L., Ignition in mixtures of coal dust, air, and methane from abrasive impacts of hard materials with pneumatic pipeline steel, U.S. Bureau of Mines R.I. 8201 (1976).

Kemenetski, F., *Zh. Fiz. Khim.* **13**, 738 (1939).

Kim, Ann. G., Laboratory studies of the spontaneous combustion of coal, U.S. Bureau of Mines J.C. 8756, Pittsburg (1977).

Kingman, F. E. T., Coleman, E. H., and Rogowski, Z. W., The ignition of flammable gases by sparks from aluminum paint and rusty steel, *J. Appl. Chem. Lond.* **2**, 449 (1952).

Leuschke, G., *Fifth International Fire Protection Seminar Vol 1*, pp. 22–24, Karlsruhe (1976).

Merzhanov, A. G. (1967). *Combustion and Flame* **11**, 201.

Powell, F., Ignition of gases and vapors by friction and impact, *Industr. Eng. Chem.* **6**, 29 (1969).

Rae, D., Safety in mines research report, 129 SMRE Sheffield, U.K. (1956).

Semenov, N. N., Zur theorie des verbrennungsprozesses, *Z. Phys.* **48**, 571 (1928).

Semenov, N. N., *Chemical Kinetics and Chain Reactions*, Clarendon Press, Oxford (1935).

Tortoishell, G., Intrinsically safe systems: installation and B. S. 5345, *Electrical Review* **206**(9), 27 (1980).

Tripper, C. F. H., *Oxidation and Combustion Reviews, Vol. 2*, Elsevier, New York (1967).

U.S. Bureau of Mines I.C. 8757, Laboratory studies of self-heating of coal (1977).

Chapter 3

Bleaney, B. I. and Bleaney, B., *Electricity and Magnetism*, Clarendon Press, Oxford (1957).

Cooperman, P. A., A new technique for the measurement of corona field strength and current density in electrical precipitation, *Trans. Am. Inst. Elec. Eng.* **75**, 64 (1956).

Corbett, R. P., Ph.D. Thesis, University of Southhampton (1972).

Paschen, F. Über die zum Funkenübergang in Luft, Wasserstoff und Kohlensäure bei verschiedenen Drucken erforderliche Potentialdifferenz, *Weid Ann.* p. 3769 (1889).

Pauthenier, M. M. and Moreau Nanot, M., Le charge des particules spherique dans un champ ionise, *J. Phys. Radium* **7**, 590 (1932).

Schwab, A, J., *High Voltage Measurement Techniques*, M.I.T. Press, Cambridge Mass. (1972).

White, H. J. *Industrial Electrostatic Precipitation*, Addison Wesley, Reading, Massachusetts (1963).

Chapter 4

Boschung, P., Hilgner, W., Lutgens, G., Maurer, B., and Widner, A., An experimental contribution to the question of lightning-like discharges in dust clouds, *J. Electrostatics* **3**, 303 (1977).

Cross, J. A., Electrostatic problems in powder handling, *Sixth Conference on Powder Technology, Powtech*, Institute of Chemical Engineers, Rugby, U.K. (1979).

Eckhoff, R. K., Minimum ignition energies of dust clouds, *Combustion and Flame* **24**, 53 (1975).

Eckhoff, R. K., Are long sparks more effective dust explosion initiators than short ones?, *Combustion and Flame*, **27**, 129 (1976).

Heidelberg, E., *Berufsgenossenschaft Betnebssicherbeit*, p. 265 (1960).

Heidelberg, E., Generation and ignition of brush dischargers, *Institute of Physics, (U.K.) Conference, Series 4*, 147 (1967).

Hughes, J. F., Corbett, R. P., Bright, A, W., and Bailey, A. C., Explosion hazards and diagnostic techniques associated with powder handling in large silos, *Institute of Physics Conference on Static Electrification, Series 27*, Institute of Physics, London (1975).

Maurer, B., *Ger. Chem. Eng.* **4**, 189–195 (1979).

Morris, K., 3rd Year Project, Department of Electrical Engineering Southampton University, (1978).

Secker, P., Static eliminator for difficult industrial applications, *Institute of Physics Conference on Static Electrification, Series 48*, pp. 115–123, Institute of Physics, Bristol and London, (1979).

Allen, T., *Particle Size Measurement*, Chapman and Hall, London (1968).

Brown, K. C., A review of present methods of testing industrial dust for inflammability, Safety in Mines Research Report No. 21, Sheffield, England (1951).

Craven, A. D., and Foster, M. G., Dust explosion prevention: determination of the critical oxygen concentration by the vertical tube method, *Combustion and Flame* **2**, 408 (1967).

Dorsett, H. G., Laboratory equipment and test procedures for evaluating explosability of dusts, U.S. Bureau of Mines R.I. 5624, Pittsburgh (1960).

Eckhoff, R. K., Minimum ignition energies for dust clouds, *Combustion and Flame* **24**, 53–64 (1975).

Gibson, N., Decomposition, fire and explosion hazards from dusts: a review, *Institute of Chemical Engineering Symposium, Series 3*, Rugby, U.K. (1981).

Gibson, N. and Rodgers, R. L., *Institute of Chemical Engineering Symposium, Series No. 58*, Manchester, U.K. (1980).

Grewer, T., On self-ignition of dust deposits, *Staub Reinhalt. Luft* **31** (3), 97–101 (1971).

Hub, L., Investigation of thermic hazards, *Loss Prevention Symposium, European Federation of Chemical Engineers*, Basel, Switzerland (1980).

Liebman, I., Conti, R, S., and Cashdollar, K. L., A dust cloud concentration probe, *Rev. Sci. Instrum.* **48** (10), 1314 (1977).

Pineau, J., Giltaire, M., and Dangreaux, J., *Cahiers de Notes Documentaires*, No. 83 2e trimestre, p. 33, Note 1005-83-7, Institute Nationale de Recherché et de securitié, Paris (1976).

Pope, M. H., and Judd, M. D., *Differential Thermal Analysis: A Guide to Techniques and Its Application*, Heydon, London (1977).

Townsend, D. I., Thermal hazard evaluation by accelerating rate calorimetry, *Acta. Thermochemica* **37**, 1–30 (1980).

Zacta, A. V., Application of hazard analysis in screening for chemical process hazards, *Acta. Thermochemica* **28**, 7–13 (1979).

Chapter 6

Andow, P. K., *IEEE Trans. Reliability* **29** (1), 2 (1980).

Andow, P. K., Lees, F. P., and Murphy, C. P., The propagation of faults in process plants: a state of the art review, European Federation of Chemical Engineers Conference on Chemical Process Hazards VII, *Institute of Chemical Engineering (U.K.) Symposium, Series 58*, 236, Rugby, U.K. (1980).

Fussell, J. B. and Rovers, G. J., Fault trees, *IEEE Trans. Reliability* **33**(4), 57 (1974).

Gibson, N., in European Federation of Chemical Engineering Conference, *Institute of Chemical Engineering (U.K.) Symposium, Series 25* (1968).

Gibson, N., Decomposition, fire and explosion hazards from dusts—a review, *European Federation of Chemical Engineers Symposium, Series 63*, 1 (1981).

Gibson, S. B., *Symposium on Process Industry Hazards*, Aston University, Paper A3, Birmingham, England (unpublished).

Hoenig, S. A., in *International Symposium on Grain Elevator Explosions*, National Academy of Sciences, Washington, D.C. (1978).

Kletz, T. A., Loss prevention and safety promotion in the process industries, *Third International Symposium of European Federation of Chemical Engineers*, Society of Chemical Industries, Basel, Switzerland (1980).

Lawley, H. G., Operability studies and hazard analysis, *Chem. Eng. Prog.* **70**, (4), 45 (1974).

Lawson, D. and Simms, D., The ignition of wood by radiation, *Br. J. Appl. Phys.* **3**(9), 288 (1952).

Lees, F. P., *Loss Prevention in the Process Industries*, Butterworth, Washington, D.C. (1980).

Stoll, A. M. and Chianta, M. A. *Trans. N.Y. Acad. Sci.*, pp. 649–670 (1971).

Chapter 7

Bartknecht, W., *Forschungsbericht F45*, Bundesinstitut fur Arbeits Schutz (1971).

Bartknecht, W., Explosion pressure relief, AIChE *Loss Prevention Symposium*, pp. 93–105 (1977).

Bartknecht, W., Report of investigations on the problem of pressure relief at explosions of combustible dusts in vessels, *Staub Reinhalt. Luft* **34** (11), 381 (1974).

Bartknecht, W., *Explosions* (Translated by Burg and Almond) Springer Verlag, New York (1981).

Chappell, W, G., Pressure-time diagram for explosion vented space, *AICE Loss Prevention Symposium*, pp. 76–86, Institute of Chemical Engineers, New York (1977).

Cocks, R. E., Dust explosion prevention and control, *Chem. Eng.*, November (1979).

Donat, C., Pressure relief as used in explosion protection, *AIChE Loss Prevention Symposium*, Institute of Chemical Engineers, New York (1977).

Donat, C., Application of explosion pressure relief, *AIChE Loss Prevention*, CEP, Vol. 11, Part 1, pp. 93–105 (1977).

Dorsett, H. G., Murray, M., Nagy, J., and Rogers, R. P., Laboratory equipment and test procedures for evaluating the explosability of dusts, U.S. Bureau of Mines R.I. 5624, Pittsburgh, (1960).

Dorsett, H. G., Jacobson, H., Nagy, J., and Williams, R. P., U.S. Bureau of Mines R.I. 5624 (1960).

Dust explosions in factories, SHN 830, Department of Employment and Productivity, H. M. Factory Inspectorate, H.M.S.O., London (1968).

Dust explosions—analysis and control, Factory Insurance Association Hartford, Connecticut.

Dust explosions in factories, Health and Safety Executive, Health and Safety at Work No. 22, H.M.S.O. London (1975).

Hartman, I. Cooper, A.R., and Jacobson, M., Recent studies on the explosability of corn starch, U.S. Bureau of Mines R.I. 4725, Pittsburgh (1950).

Howard, W. B. and Karabinis, A. H., Loss prevention and safety promotion in the process industries, *Third International Symposium, Basel, Switzerland*, Society of Chemical Industries, Basel (1980).

Maisey, H. R., Gaseous and dust explosions venting, *Chem. Proc. Eng.* October, 527 (1965).

Moore, P., Characterization of dust explosability: a comparative study of test methods, *Chem. Indus.* **7**, 430–433 (1979).

Moore, P. E., Dust explosion hazard assessment, *International Chemical Engineering Symposium, Series No. 58*, 77–88, Institute of Chemical Engineers, Rugby, U.K. (1980).

NFPA Standard No. 68, Explosion venting guide, National Fire Protection Association, Boston (1978).

Nagy, J., Zeilinger, J. E., and Hartmann, I., Pressure relieving capacities of diaphragms and other devices for venting dust explosions, U.S. Bureau of Mines R.I. 4636, Pittsburgh (1950).

Palmer, K. N., The relief venting of dust explosions in process plants, *Int. Chem. Eng. Symp. Ser. 34*, 142–147, Institute of Chemical Engineers, Rugby, U.K. (1971).

Palmer, K. N., Dust explosion venting—a reassessment of the data, Fire Research Note No. 830, Fire Research Station, Boreham Wood, U.K. August (1970).

Palmer, K. N., *Dust Explosions and Fires*, Chapman and Hall, London (1973).

Peterson, P., *AIChE Petroleum Exposition, February, Houston, Texas*, American Institute of Chemical Engineers, New York (1967).

Pineau, M. M., Giltaire, M., and Dangreaux, J., *Cahiers de Notes Documentaires*, 1 er trimestre, Institute Nationale de Recherché et de Securité, Paris (1978).

Raferty, M., Explosability tests for industrial dusts, Fire Research Technical Paper No. 21, Fire Research Station, Boreham Wood, U.K. (1968).

Rasbash, D. J., and Rogowski, Z.W., Gaseous explosions in vented ducts, *Combustion and Flame* **4**, 301–312 (1966).

Runes, E., *Loss Prev.* **6a**, CEP Technical Manual, American Institute of Chemical Engineers, New York, pp. 63–67 (1972).

Rust, E. A., Explosion venting for low pressure equipment, *Chem. Eng.*, p. 102, November (1979).

Schwab, R. F. and Othmer, D. F., Dust explosions, *Chem. Proc. Eng.*, pp. 165–174 April (1964).

Thomas, F. E., *Explosion Venting as a Means of Controlling Dust Explosions*, Weyerhauser Co., Tacoma, Washington.

Tonkin, P. S., and Berelemont, C. F., Dust explosions in a large scale cyclone plant, Fire Research Note No. 942, Fire Research Station, Boreham Wood, England July (1972).

Index

GPSR Compliance

The European Union's (EU) General Product Safety Regulation (GPSR) is a set of rules that requires consumer products to be safe and our obligations to ensure this.

If you have any concerns about our products, you can contact us on ProductSafety@springernature.com

In case Publisher is established outside the EU, the EU authorized representative is:

Springer Nature Customer Service Center GmbH
Europaplatz 3
69115 Heidelberg, Germany

Batch number: 10620163

Printed by Printforce, the Netherlands